THE ANSWER LIES IN THE SOIL

Peter Binns

MINERVA PRESS
LONDON
MIAMI DELHI SYDNEY

THE ANSWER LIES IN THE SOIL
Copyright © Peter Binns 2000

All Rights Reserved

No part of this book may be reproduced in any form
by photocopying or by electronic or mechanical means,
including information storage or retrieval system,
without permission in writing from both the copyright
owner and the publisher of this book.

ISBN 0 75411 134 2

First Published 2000 by
MINERVA PRESS
315–317 Regent Street
London W1R 7YB

Printed in Great Britain for Minerva Press

THE ANSWER LIES
IN THE SOIL

*To Judy
without whom
this would never
have happened.*

Chapter One

The thick fog hung over the land like a shroud over a reluctant corpse. It was dense and sulphurous with that dead taste that today's youth would not recognise, thanks to the clean air act. It was called smog.

I opened the front door for the fourth time. Where was she? My imagination ran riot. Was she lost? After all, we had only been living here a few months since we were married. But all I could see was the curling fog.

Suddenly, in the total quiet that always seems to accompany fog, a funny putt-puttering broke the silence, and a man bid my wife 'goodbye' from one of those bubble cars with a front opening door.

Totally forgetting my worries, I asked jealously, 'Who was that?' This was my first glimpse of the man who was to become our friend and business partner, Clifford Potts.

I had met my wife, Judy, at a party of a mutual friend of ours and discovered we both worked for the same City Authority; she for the Health Department and I for the Housing Architects. We seemed to hit it off right away.

Eighteen months later, on taking over the rent of a house, we decided that, although we had decided to get married the following September, it would be silly to pay rent on an empty property for nine months and not live in it. So we got married.

Having only a few days holiday left, the only time we could take a full week off was at Christmas, by adding in the statutory holidays. We married on a Saturday, on Christmas Eve. The church was obviously decked up beautifully, and all for free!

My best man, Alan, turned out to be ten times more nervous than I was, so I spent all my time reassuring him and forgot all about my own worries. I literally had to drag him up when the Wedding March started.

Already portents were gathering. On coming out of the

church, we were met by such a downpour that no photographs were possible so we were whisked away to the reception at a local pub with the unlikely nickname of The Pig Muck, hardly an auspicious start to married life.

Due to the rush and the difficulty of booking somewhere at that time of the year, my parents had persuaded a friend of theirs to open up their boarding house at Southport, just for us. So, following a party held at my parents' house in the evening for all those friends who could not fit into the tiny pub, we set off, after pretending we would be taking my wife's parents home first. It fooled everybody, because, half an hour later, when the usual crowd set out to tie shoes and tin cans, together with an old kipper onto the exhaust, we had already made our escape.

The journey to Southport was about eighty miles and we arrived around eleven at night to be met by another party in full swing.

Near midnight, Judy pleaded tiredness and ten minutes later to knowing winks and nudges, I retired too. It was all we could do to fall into bed and we were fast asleep before our heads touched the pillow; hardly romantic and nothing like the folks downstairs were imagining.

Next day after breakfast, we wandered, hand in hand down Lord Street, the main shopping street, feeling as if we had placards ten feet high proclaiming 'Just Married' on our backs.

A lady approached. Oh no! We had been spotted. She was replete with pen and clipboard and announced that she was from the BBC. Fame at last!

'Could you,' she asked, 'tell me what programmes you watched on television last night?'

Panic. What should we say? Blushing we explained that we had just got married the day before and we had not stayed up watching television. It was her turn to go red, and wishing us well, she dashed off, probably to down a strong whisky to steady herself to face the next impossible reply.

Southport being Southport, virtually everything was closed down for the winter. However, we did manage to find one show on; an Ice Show! in the middle of winter? Shivering in coats, jumpers and scarves, we were determined to enjoy ourselves and prove our love could keep us warm. Some hope!

Back home, another problem had occurred. The people in our new house had not been able to move out, as the people in the house they had to move to could not vacate until their new property had been finished. So we had to spend an uncomfortable month at my wife's parents' home, sleeping in a three-quarter bed, which was cosy, to say the least!

Finally, we moved into our new home, on bare boards, with the usual packing cases for table and chairs. This wasn't because we lacked furniture, but because the whole house needed decorating and the less clutter in the way the better.

The property had originally been part of a larger house. A new semi had been created from a store above a workroom with a built-on kitchen and bathroom above, together with two rooms taken off the house next door. Our landlord lived next door.

The former house and workshop had belonged to a hatter, (not the Mad Hatter, the felt kind). The area was well known for producing felt hats in their heyday, first being produced as a cottage industry.

There was a small, raised garden at the back, consisting of a lawn surrounded by a border on three sides, with the usual perennials, shrubs and a few roses. All we really did was tickle it and keep it tidy, not what you would call real gardening. We did not realise what fate had in store for us at this stage. If only we had known…

Why tell you about the past history of the house? It has a bearing on two episodes in our life. The first of these happened in the first few weeks, as we were clearing up and starting to decorate.

The place was in a right mess; three layers of carpet on top of one another used as underfelt and seven, yes seven, layers of wallpaper on the walls. How did we know? It was easy to count them because the former occupants had papered round the furniture, leaving a perfect image of the grandfather clock and dresser, ghosts of the originals.

Across the centre of the living room was a beam and we decided we would like to have horse brasses along this. But there was a central light fitting, so we opted instead to have lights either side.

Now came the problem. Try as I might, I could not cut into the ceiling. It turned out that under the plaster the whole thing was slate tiled. The hatter had used steam to cure the hats and then stored them in the room above. To prevent the steam from below getting into the storeroom above, they had used a slate ceiling. You learn something everyday.

The other episode was much more hair-raising. On going to bed one night, I found it difficult to get to sleep. I felt that there was somebody in the room and try as I might, I could not get the thought out of my head. Not wanting to frighten Judy, I strained my eyes to see if there was anything there.

Suddenly, Judy whispered, 'I don't want to appear silly, but I am sure that somebody is in the room.'

'Where?' I said.

'Next to the fireplace in the corner', she replied.

I quickly switched on the light, but there was no sign of anybody. I admitted that I had had the same feeling and had sensed the 'person' in the same place. Obviously, we must have been mistaken and eventually, we went to sleep.

Next day, meeting our landlord, we jokingly mentioned the phenomenon expecting him to laugh with us. Instead, he said, 'So you've met our ghost, have you?' This came from a person we would normally have considered a tough nut.

'Where have you put the bed?' he asked. So I told him.

'That's where the chap died in bed with rain pouring through a hole in the roof. Change it to the front wall and you should have no more problems.'

So we did, and we never had a moment's problem from then on. Where we had imagined the person to be was where there had once been a door between the two houses. Some believe it, others don't.

Chapter Two

Following the night of the fog, Cliff became a regular visitor, ferrying Judy and her friend to and from work as he also worked at Manchester Town Hall. I had moved on to a post with Macclesfield Borough.

When Judy found out that Cliff was a bachelor living on his own and existing on the usual baked beans on toast and bacon and egg, she invited him to have a meal with us.

Afterwards, we wandered round the small back garden. Cliff was fascinated, commenting on the flowers in the borders and suggesting ways to make it better. It turned out that he only had a paved back yard and a very small front garden. He adored gardening, but did not have the space to do anything.

He had trained in Horticulture at Manchester University and passed with an Honours Degree, whilst working straight from school in the local Parks Department. His goal had been to reach Parks Superintendent, but this had not been possible. It was just after the war and those jobs were already filled with young men and it was a dead man's shoes position. So he had changed jobs, entering Manchester City council as a trainee Public Health Officer. In those days they were called Sanitary Inspectors, but what's in a name?

He had qualified at what is now Salford University and had worked for a number of years in the Health Department. Rather set in his ways, we must have been quite a cultural shock for him. Judy, being a total extrovert, really livened things up. You never knew quite what to expect from her next.

A typical example was the time when we were decorating the living room and painting the ceiling, having toiled for weeks to clear up the grime and dirt. (The former tenants had used the space under the bottom drawer of a cupboard in the living room to empty the ashes from the fire and an old gas clothes boiler had so much green slime behind it that it was a wonder that it did not get up and walk.)

We were each up a stepladder, brushes in hand, tired and irritable. A chance remark from me, not expressed very well, resulted in an argument, still standing on our ladders. Judy lost her temper and sloshed my head with a full brush of pale yellow emulsion. Dripping with paint and a little shocked, I reciprocated. In no time at all, we were battering at each other with yellow paint, for all the world like two knights of old in their respective castles, firing away at each other with boiling oil.

Soon we started to see the funny side of it. Howling with laughter, with tears in our eyes, paint mingling with our hair and running down our faces, we descended the ladders to be met with a straight faced, but puzzled Cliff, who had come to give us a hand. Macclesfield nuthouse here we come! (For those that do not know, there is a mental hospital there, or possibly not any more, so many changes have taken place in the Health Service.)

Cliff became a regular visitor to our house, staying for meals and an evening's entertainment. We had a television, which we had obtained from a relative from the Midlands. It was tuned to the local stations in the Midlands, and try as we might, we never were able to get it tuned to BBC in the North, and all we had was ITV. Despite this, we were not able to get out of paying a licence fee. In any case, the programmes were not very much to our taste, so we used to play cards a lot.

In the winter of 1962, there was an extremely persistent cold spell, which froze the ground to some depth. We had noticed a strong smell of gas, so we rang the Gas Board. They said they would send somebody round.

The next evening, there was a knock at the door. I opened it to a large gentleman with an even bigger hooter, smoking a pipe. He was from the Gas Board, could he come in and investigate? Still puffing away on that pipe, he proceeded, in between puffs, to smell the air with his enormous nose.
Now if there had been a concentration of gas, his pipe would probably have blown us all to kingdom come. Luckily, it was spread fairly uniformly throughout the house, but how was he to know that?

He said that he would report it.

The next day, they turned up with drills, and set about digging holes alongside the house to disperse the gas, but it seemed to make little difference.

They then drilled holes in the road to try to find where it was coming from, to no avail. They went away to consider their next move, meanwhile we were being slowly gassed.

Incidentally, no one suggested that we should stop lighting an open fire in the sitting room, and seeing as it was severely cold and the man with the pipe had not blown us up, we continued to have a fire. Who were the fools, was it them or was it us?

The next night, we sat playing the inevitable game of cards. Outside, it was another of those pea-soup fogs. Again, there was a knock at the door. Opening it, I was met by a man in a huge gas mask, looking just like an alien.

'Wuj oo moo hut?' he said.

I indicated that I could not understand him.

He removed his mask. 'Would you move out?'

'When, now?' I said.

'Yes immediately. Is there somewhere you could go to?' he said, this at half past eleven at night. Luckily, Cliff was there.

Hastily grabbing eiderdowns, pyjamas and a hot water bottle, (remember it was very cold), we drove off in Cliff's bubble car towards his house. This was not easy, what with three of us plus accoutrements, the windows misting up and the fog outside. Cliff did his best, trying to keep the lampposts on his left side, but then we passed a side road, which meant a large gap in the lampposts. Cliff saw the next one, so keeping it to his left, he carried on.

'Stop!' I shouted.

'Why?' he enquired.

He couldn't see what I could see; shop windows, immediately to my right.

Now I have to explain that these bubble cars were imported from Germany, and being so small, it was not thought necessary to change the drive to the right, so Cliff was on the left, with Judy in the middle and I was on the right. We had stopped because we were driving on the pavement. Having screeched to a halt, we had to decide what to do next.

Cliff produced a torch and Judy got out clutching it and the hot water bottle, and set off down what she hoped was the road. Cliff drove, while I checked that we were not on the wrong side of the road again.

We finally made it, and spent a rather uncomfortable night, all the while wondering what was happening to our home. Would it still be there when they had finished? We had no idea.

After two days of this, constantly checking to see whether we could return, we were informed that everything was all right, and we could go back.

It turned out that they had all assumed that the trouble was in the old gas pipe, never imagining that a brand new high-pressure gas main, only put in two years previously, was the culprit. It had taken them a week to find it. That gas smell took a long time to dissipate, but at least the cold spell was over.

Looking back, I can still see in my mind's eye the man with the huge nose and pipe, and the man with the gas mask and indecipherable language. I am sure that they did not realise how ridiculous they looked.

Chapter Three

As a married couple, we had the usual arguments. After a rip-roaring one, about what, for the life of me I cannot remember, I stormed out of the house. We lived not far away from my parents' house, where we kept our car, so I went and sat in the car to calm down.

Meanwhile, Judy, regretting the incident, set out to find me and thinking I had gone for a walk, went on 'The Low', which, in the daft parts we come from, is actually a hill. It was pitch dark with no lights and feeling rather frightened, she called my name, but the only answer she got was the mooing of a cow, not something to calm the nerves.

At the same time, also regretting what had happened, I returned home, to find Judy gone. The only thing I could think of was that she had gone home to her Mum. So I shot back to the car and drove the eight miles to her parents' home, cursing myself for all the trouble.

On arrival, I sheepishly told her parents what had happened, but there was no sign of Judy. Perhaps she was still on the bus and I had overtaken her.

After half an hour, I was just about to return home when the phone rang. It was Judy. Having returned home through the darkness amidst even more disconcerting noises, thoroughly frightened and confused, she was ringing her Mum to ask what to do.

Imagine her surprise to find out I was there looking for her. When I returned home, we laughed and giggled at our respective attempts to find each other, the argument totally forgotten.

There were other moments, like the time she locked me out. Hammering on the door, I insisted on being let in. There was no answer, so I opened the letterbox, and peered in. That was a mistake. She was waiting for me, and immediately spat straight in my eye. (Well, she is rather volatile!) She regretted it afterwards and let me in.

One other time, just before our first Christmas in the house, she decided that I should be a Father Christmas. She proceeded to splatter my beard with flour. I ran outside the back door, with her chasing, intending more harm to me, only to be met by our landlord, who, shaking his head, remarked, 'What are children coming to these days?' At least it made Judy back off.

Chapter Four

We had not bought a house in the first place for two reasons. One was that I was still training in architecture (when we married in fact Judy was earning more than me). Secondly, because I was not qualified, I had to change jobs frequently in order to advance. That meant that I had to be capable of moving house as well. But, now that we had such a good friend as Cliff, we considered exploring a joint venture, which we could embark on, but for the time being, keep our respective full-time jobs.

I had always looked for additional work to earn more money. Judy felt no sense of adventure in a rather boring clerical job and Cliff really needed pulling out of his complacent bachelor life. The answer, we decided, lay in our common interest in gardening. So we decided to set up a garden centre.

But, where and how? Working for local government, Cliff and I were both aware of certain trends. Cliff knew that Manchester was going to build overspills to the east, including one in a small town called Glossop and I could see that residential development, having built up north, west and south of Manchester, was also moving eastwards. Since the Pennines were in the way, Glossop was the only place where such development could culminate.

Glossop was a rather depressing, run-down, cotton town, and people were leaving it in droves, not coming into it. The local cotton industry had virtually collapsed and another big employer, paper production, seemed doomed due to the poor transport and infrastructure. Compared to other towns near motorways and large railway networks, Glossop was a dead end. It was situated at the bottom of the notorious Snake Pass to Sheffield and side-tracked off the Manchester to Sheffield Railway, which passed through the Woodhead Tunnel, by a spur line to Glossop Station.

We saw it as a challenge. We needed a fair-sized piece of ground, on or near a main road, with a house of course, to the east of Ashton-under-Lyne. And it needed to be cheap! The only type of property to offer all that was a large Victorian house with grounds. Luckily, prices were low in the area and even lower for this type of property, so we began the hunt.

After many interesting but fruitless searches over weekends, we found a large Victorian house with three and a half acres, in a village, not far from Glossop, on a bumpy track, just off the main road over Woodhead to Sheffield. It had not been lived in for three years and its last use had been as a club, which had gone bust, but it was cheap.

Although cheap, it may as well have been a gold mine; no building society or insurance company would advance us any money on it. We had next to nothing to put towards it ourselves, let alone the money we would need to set up the nursery.

We tried everything. Finally through a friend of a friend, we made an approach to a financier, who, they claimed might be able to put up the money for the business, if the circumstances were right.

Contact was made and he suggested that we meet at a very upmarket restaurant, not far from the site we were interested in. We could not refuse, but could we afford it? We had to make a good impression, so counting our pennies, we agreed.

Judy was cowardly and refused to go. She said she would have nothing to contribute and anyway, an extra mouth would only cost even more. So just two of us prepared to meet our hoped-for benefactor.

The appointed day arrived and in best 'bib and tucker', we set off to meet our patron to be (or not to be that was the question). We had arranged to meet in the car park of the restaurant and arriving early, we had to wait some anxious moments for him to arrive, but he finally came and we went in for lunch.

We had a drink whilst we looked at the menu. Our worst fears were confirmed when we saw the prices. I looked at Cliff and he at me. It was going to cost us a week's wages and I was not at all sure we had enough between us. I mentally checked what I had. It was not even half enough at those prices. I hoped Cliff had enough to make it up.

Totally unconcerned, Mr 'X' ordered one of the most expensive meals, recommending we had this or that, which were almost as much and then asking for a bottle of wine which on its own, was half my week's wage. I had my chequebook but, mind you, there was next to nothing in my bank account. I had visions of doing the washing up for a week, not to mention the fear that, when he found out how broke we were, nothing would induce him to loan any money to paupers such as us.

I have never eaten a meal before or since, in such abject terror. All the while I was attempting to appear interested in what he had to say and show some enthusiasm in what we intended to do. It was terrifying. After the meal he ordered liqueurs and lit up an expensive cigar. Oh hell! What would this cost us? When the bill came, I was trembling and neither Cliff nor I could bring ourselves to pick it up (it would certainly bite!)

Suddenly, the most amazing thing happened. Mr 'X' picked up the bill and drew a well-filled wallet out of his inside pocket. Both Cliff and I, although desperate that we should not have to pay the bill, immediately clamoured to be allowed to cover it, as he was our guest.

Perhaps he could see past all our bravado. He insisted on paying, as he had chosen the restaurant and in any case, since he had found us such good companions, it was worth it to him.

If I had uttered a sigh of relief, it would have been heard in London. That is what I call true generosity and I will remember him and that act all my life.

We showed him the property and explained how we would carry out plans to create a garden centre. He listened avidly to what we were saying and then dropped the bombshell. He offered us twice as much as we wanted, explaining that we would need the extra capital to create the profits quickly, as he wanted us to pay him back in five years with interest. We were staggered, we had no answer. All we could do was thank him. We said we would think about his proposition and let him know in a few days time.

What were we to do? We had the money, more than we wanted, but to guarantee to pay him back with interest in just five years, was really not on. We were adventurous and prepared to take a chance, but not that adventurous. After all, the sum was twice what we jointly earned in a year and if we failed, we would be homeless and broke. At least at the moment we were just broke.

I have this theory; if you have money, unless you are an idiot, you can make money. Similarly if you have no money, no home and no job, you have a fair chance of making money, if you fail, you can only fall to the bottom where you started. But, if you are part-way up, you always stand to lose more than you had.

Both Cliff and I were in the part-way up category and we felt that taking this particular plunge was likely to put us further down the rungs of the ladder than we were. So, reluctantly, we had to turn down the offer. But we hated ourselves for doing it; one, because we were refusing an opportunity and two, because he had paid for that meal. What to do next? Properties like that were not ten-a-penny and to find that had taken weeks.

Chapter Five

We were really down in the dumps. It must have shown because colleagues at Stockport, where I now worked, asked me what was wrong and I had to tell them the whole bitter tale.

Now fate took over. One of those listening to the sorry tale lived in a village called Gamesley, part of the Borough of Glossop. He informed me that there was a large property in the village that was soon to be up for sale. The cleaner there had heard the owner and his wife talking about it and the village, being a village, soon had the gossip out.

Trying to work up some enthusiasm, I went home and reported the facts to Judy and Cliff, believing that any new attempt would be doomed to failure, due to cost or an inability to find the money. I had reckoned without Judy's exuberant character. Ever since we had discussed the idea, it had bloomed in her mind and loomed large in her interest for the future and now there was another chance, there was no stopping her. Before we knew where we were, Judy had us speeding round that very evening, to see what it was like.

The house was typically stone-built Victorian, Georgian in style, standing back from a main road, surrounded by a six foot high stone wall, an eight foot high privet hedge and eighty foot high trees. Next door was a public house and in between them ran a road down to a railway goods yard. It faced onto a recreation ground, with the village properties running up and down the main road and a lane opposite the public house.

The land was at the back of the house, approximately two acres, with a high brick wall surrounding two sides and facing almost directly south, a perfect suntrap. The land was also lower than the side road, terraced down from the house. Adjacent to this side road, was a greenhouse some one hundred feet long. Additionally, on the other side, there were garages and other sundry buildings.

Apart from the land not being visible from the main road, it was perfect and the road to the front was a through road, which would encourage passing trade.

What next? Wait for the advert? Judy would have none of it.

'Go and ask,' she said.

So Cliff and I approached the front door, which, perversely, was on the side, and rang the bell with some trepidation.

The gentleman of the house answered. We told him that local gossip had it that they were considering selling and that we would be interested if the price was right. He confirmed that they were thinking about it, but said that they had not made up their minds. He took our names and telephone number and agreed to contact us if he was selling and to give us first offer.

We got back to the car. Judy was ecstatic when we told her and immediately started making plans. We had to calm her down, pointing out that the owner had not said he was going to sell.

Doom and gloom, but spiced with anticipation, settled in. Once again we seemed to be thwarted.

Three whole months went by with nary a whisper, then, suddenly, contact! Yes, he was interested in selling and would we like to inspect and talk about the price? Would we? We arranged a meeting that instant.

Now doubts appeared. Could we afford it? After all, unlike the last property, this one had been occupied since it was built. We could only go and see.

Judy was so unnerved by it all that she would not go in, but sat, once again, in the car round the corner.

Cliff and I were shown round. The house was huge; six bedrooms, four reception rooms, a huge kitchen with scullery and pantry. It was far bigger than we needed. The hall and landing alone was the size of a modern three-bed semi. But the gardens with the greenhouses, garages and outhouses were just what we needed.

On top of this there were two adjacent cottages, both rented, although one still had only gas lighting. Having had some experience in the past with rented properties and the repairs these needed, we could see that any capital needed to set up the business would be swallowed up in repair costs on the cottages.

We suggested that he might sell without the cottages. In hindsight, this turned out to be a bad move.

After some thought, he agreed and named a price, which was below what we were expecting. With glances and nods, Cliff and I went along with this and a bargain was struck. We explained the problems we had experienced before on a mortgage, but he had an insurance mortgage and he was convinced that he could get them to agree to give us the necessary funds.

We went home full of ideas and plans, cracked open a bottle of sherry (we are not in the champagne class) and talked and talked until after midnight. This time we might make it.

Chapter Six

A week later, with the insurance company agreeing to give us a sixty percent mortgage, we looked at ways and means of finding the remainder. Cliff had his house, which he put up for sale and I had an equivalent amount, from selling a piece of land I owned. Doing our sums, this left us with just two hundred and fifty pounds to start the business, hardly a fortune and we still had not allowed for solicitor's fees.

Here we were lucky. My father was a solicitor, although he did not deal with house conveyancing. When we approached him with the proposition, he nearly had kittens. What did we think we were doing putting such a potential millstone round our necks? But when we pointed out all the plus points and the fact that we were paying only around the price of a good three-bed house, he agreed to do the conveyancing for free, but he still thought we were mad. Being a conscientious parent and a typical solicitor, he wanted to see the place, so we arranged to go round the next weekend.

When we arrived, he was even more staggered, having not anticipated its size. Then he became the practical man and devil's advocate. What about the carpets, curtains and furniture? We had totally overlooked these in our euphoria. He struck a bargain with the owner, whereby he would do the conveyancing for both sides free of charge. (This is now illegal, I don't know whether it was then but it is too late to object now.) He would do this in exchange for the carpets, a few of the curtains and some of the larger items of furniture, which had originally come with the property and been included in the sale to the present owner.

Then the next bombshell fell with a resounding thud. The insurance company decided to withhold three hundred pounds of the mortgage money, until the outside of the house had been painted. It transpired that the present owner had reneged on a deal to paint the house when he had obtained a mortgage from the insurance company, and they were not going to get caught out again.

We persuaded the owner that he should stand the delay in this payment, as it was his fault that we were put in this predicament. Faced with the possibility of us backing out of the deal, he capitulated. We, for our part, agreed to do the painting within three months of occupation.

The next episode is something that I will never live down. My father was somewhat old fashioned and insisted that the money we were providing for the purchase, should be handed over in cash. In my case this amounted to a thousand pounds and I duly arranged with the bank to make the money available.

Never having had to deal with this kind of large sum, and in those days an Austin Mini cost only five hundred and sixty pounds, I arrived at the bank with a briefcase. Imagine my composure, when I was handed a foolscap envelope and had to pretend my empty briefcase was heavy, as I pocketed the envelope and walked away. Was my face red! It was not helped when I reached home and told Judy, who collapsed laughing.

Cliff had sold his house and so the total amount was now available, leaving us the princely sum of two hundred and fifty pounds to set up the business. Full of euphoria and hope, we could not see that this was hardly enough.

The deal was settled and a date set for possession. Next came the arrangements for the removal. You should have seen the looks we got when we told them we were moving two houses' furniture into one. Suddenly, we realised that everyone thought we were setting up a *menage a trois*, which was the furthest thing from our minds. There were some tricky explanations to give before most people believed us. Well, they could believe what they wanted, we were all set to go ahead and nothing was going to stop us now!

We arranged to make a final inspection, to see what we needed or had forgotten. The thing we had left out soon struck us – the curtains. Apart from the ones in the lounge, the rest were shot. We measured up, but with sixteen large eight-foot by four-foot windows to cater for, the material alone came to around a hundred and twenty pounds, almost three weeks wages.

Those two hundred and fifty pounds was becoming smaller by

the minute. Luckily, a friend agreed to run up the curtains at no charge, give or take a drink or two.

What else had we not foreseen? We decided it was too late now and whatever it was would have to wait until we could afford or deal with it.

The move took place on my birthday, 25 February. Luckily it was not a Friday as well, but it may as well have been; you know what they say about moving on a Friday – it was dark, damp and dismal.

We dropped Judy off at the house, which when viewed in that light, was more like a gothic horror movie set, only the lightning was missing. We went off to Cliff's house, loaded up, then on to our house to do the same and finally off to our new home bathed in its ethereal mist.

Meanwhile, Judy wondered what on earth we had done. Wandering around a huge empty Victorian house, surrounded by stark, lifeless trees in the gloom of winter was not something to set you dancing with joy, and she was extremely pleased when we arrived back. She was soon so busy rushing around organising where things went that she had no time to consider the gloom.

Later, over a cup of tea, we realised we had made it, we were on our way. It gave us a warm feeling that dispelled the murk round us. Little did we know what the future held for us, but for now, our dream had arrived. It was 1964, the optimistic years were ahead of us and our garden centre, which was called 'Gamesley Nursery', was born.

Chapter Seven

As the dawn broke we felt both exhilaration and fright. Had we done the right thing? But, first things first, we had to arrange the painting of the house exterior.

This brought our next shock. Even in 1964 the lowest quote we got was for eight hundred pounds, which was out of the question. It would put us into the red even before we had started to buy stock to sell and growing stock on would not happen overnight.

Well, there was nothing else for it, but to do it ourselves. So we bought an aluminium ladder, brushes and paint, all for the princely sum of thirty pounds. How did the decorators justify the remaining seven hundred and seventy pounds? We soon found out.

At the first sign of dry weather, we propped the ladder up against the gutter and Cliff set off, with some trepidation, to climb skywards, paintbrush and tin in hand.

Now, aluminium ladders are not the sturdiest of beasts and tend to bend. At about the halfway stage, Cliff felt the ladder striving to make acquaintance with the wall and in a fit of nerves, started to laugh. This was not a good thing to do, as he discovered, because this produced a slight bounce in the ladder. By this time we were joining in the laughter, which of course, made Cliff laugh even more, accentuating the bounce and making it impossible for him to continue upwards. Cowardice took over and he retreated back down in some disorder, vowing, 'never to climb that bloody ladder again!'

So, the task became mine. Now we discovered the other problem. The overhang of the fascia was some two feet, so that if you rested the ladder on the gutter, you could only paint it and the fascia, but could not reach back under to paint the soffit of the eaves back to the wall. This meant that it was necessary to paint the gutter and fascia first, then climb down, reposition the ladder against the wall, climb up again and paint the soffit. Now we knew why the quote was so high! There was over two hundred feet of it, not counting the sixteen windows, two large doors and the ironwork to the balconies over the lounge bay and the entrance porch.

Thank goodness they had not included the hundred feet of greenhouse, the two garages and outhouses; we would never have finished. As it was, it took the better part of three months' labour fitted in between poor weather and our day jobs.

We only just made it within the deadline, but the job was done, the final monies were paid and it was all ours. Well one third was, the rest was the insurance company's.

In between all this, other things were happening. They do, don't they? Joys of spring and all that. March came and Judy was confirmed pregnant. Ye Gods! What a time! She was our mainstay. Could she cope? Judy, as usual, played the whole thing down and announced it would be no problem. If only!

A friend of ours decided that we or principally Judy needed a dog to guard the house. Cliff, who had just had the unenviable task of re-housing a slum clearance area in Ardwick, had befriended an eccentric old lady who bred labradors. She kept them all in the house with her and she really lived with them, because they had the complete run of the place. Cliff had to see if he could find owners for a large number of pups and inevitably, we became the recipient of one, a beautiful golden labrador called Sheba.

However, the dogs, over a period of time, had become inbred, and Sheba was rather timid and lost. Cliff had been unable to find homes for all the dogs, and it looked as though several would have to be put down. We decided that we would take on a second dog, another bitch called Neta. But if anything, she was more neurotic than Sheba, although together they seemed to be a steadying influence on each other. They became our watchdogs.

Apart from the dogs, we desperately needed some form of transport for goods and chattels. We owned two cars, which were not designed to transport anything but people. One would have to go. I drew the short straw.

Whilst Cliff had an almost new Fiat 500, which was very frugal, mine drank petrol. It was no contest. I had to swap my beloved Sunbeam Rapier for a Mini van. Mind you, it seemed to me that the only one to get a good deal was the garage, but beggars cannot be choosers – at least beggars of our calibre couldn't.

We had one bit of luck. We were able to employ a retired farmer part-time. He was called Stanley and lived in the village, and he was a dab hand with a spade. He was faster than Cliff, and that was what we needed; the ground had not been worked for a long time. Showing him around the place, he noticed something that we, blind as we were to any disadvantages to our grand scheme, had not seen.

The previous idiots who had inhabited the house had not used the greenhouse heating, but had still left the water in. This had had the inevitable result that the first frost that they had encountered had burst the pipes, leaving both the boiler and pipes useless.

Ah well! We had not intended to heat the greenhouse at this stage, but it would have been useful for the future. (Future? What was that? We could not even contemplate it.)

Talking about heating, we were already aware before we bought the property that the previous owners had not used the heating system for the house either. It was a huge 'Robin Hood' boiler, into which four people could have easily fitted. Situated in an outhouse, feeding into the two adjacent garage buildings before crossing the drive into the house itself, via four-inch pipes and huge hospital-type radiators. On first firing it up, the previous owners had got through a ton of coke in ten days flat. They gave it up in disgust, and we had already decided to follow their noble example. We only heated the rooms we were in, but not the bedrooms. Acres of blankets and a good eiderdown or two would have to suffice.

Chapter Eight

Did I mention the billiard room before? No, I don't think I did. The place was so vast that a small room like that tends to slip your memory. (Only joking!)

There were two garages, both big enough to house four Minis. One had been built approximately forty years before, in the twenties, to house the family's first car, a Rolls Royce. This garage had an inspection pit connected to the other one by a passage. The other, an earlier building, had been a coach house with a fully fitted out billiard room above. There was a full-size table and a cupboard containing a rack of cues together with the usual score board, all tastefully set in close-boarded panelled walls and ceiling, with roof lights and suspended light over the table.

Unfortunately, this delightful feature had not been in use for a number of years and due to the lack of heating, had suffered badly from damp and mildew. Even the cushions had gone hard and the cues were warped (they would have made better bows!) We would have no time for this pursuit until several years later.

Anyway, I digress. Back to the problems at hand. We had decided to open for business in only a few weeks, at Easter, the traditional start to the gardening season.

We bought in stock and set about growing hundreds of perennials and rock plants, working late into the night. Don't forget that although Judy was gardening full-time, Cliff and I still had our respective jobs in local government. Judy's schedule also had to include the normal housework, which was a full-time job in its own right, especially with a house that big. Time was, as they say, of the essence. The only good news was the discovery of hundreds of plant pots and numerous garden canes in the storeroom adjoining the potting shed by the greenhouse: corn in Egypt!

A sign was commissioned in tasteful green and black on a white background. Well, we thought it looked tasteful. Using two old scaffolding posts hammered into the ground, with timber posts set in, the sign was bolted on and the whole arrangement placed in a commanding position against the front wall.

We were ready for business!

But what business? We soon discovered that the villagers would not venture into the grounds. After all, they had always been kept out, only being allowed on the premises during the annual garden party in days gone by. (One of the first occupants was a mill owner, who virtually owned the village and villagers heart and soul. Some ninety percent of them had originally worked for him in his mill.)

Perhaps now would be a good time to give you a potted history.

Back in the 1850s, when the Industrial Revolution was booming, Glossop became the site of a large number of cotton mills, together with associated industries like dye works, although there had been earlier mills there since the late 1700s.

In fact, Glossop as we now know it did not exist before about 1800. The area was composed of a number of small villages and hamlets, one of which was Glossop (now Old Glossop) standing to the north-east of the present centre. With the coming of the mills, which were concentrated on the valley bottom near the streams, the present town grew up around a crossroads, where the Snake Pass to the east met the main road from Chapel-en-le-Frith from the south. To the north was the other crossing of the Pennines, the Woodhead Pass. The westerly road went to Manchester.

The mills produced untold wealth for the mill owners, who built themselves grand houses to show off their new-found fortunes. One of the largest mills was Wood's Mill, named after the Wood family who set it up. One of the Woods' sons purchased the land at Gamesley in the early 1850s and built the house, which we had now bought.

But he had decided that it was not grand enough, and so he moved to an even bigger house over to the east of Glossop, called Moorfield. This was so big, that it has now been split into two houses, which by today's standards, are both still large.

The Woods were great benefactors to the town, donating

Wood's Hospital and Wood's Swimming Baths. The latter was set in Howard Park, named after Lord Howard, who owned most of the land in the area at the time. Most of the prominent buildings, parks and even the street lighting and sewage system were donated to the town by prominent families of the day. Many streets are named after them or their children.

As I said, Gamesley House was built by one of the Wood family. The first mention of a house on the site was in the deeds, when he sold it to General Booth of the Salvation Army, who used it as a convalescing and retirement home for his officers. I also understand that his parents lived there. I am not sure why General Booth decided to sell it, but he did, to the Shepleys, yet another family of mill owners, who had mills in both Glossop and the nearby town of Hyde.

The Shepleys began their first mill in the late 1700s in what is now Old Glossop and so were very well established, although perhaps not as farsighted as the later mill owners of the district who took more chances and built far bigger mills.

Mr Shepley was not a benevolent gentleman, and was noted for his slave-driving of his workers. It was rumoured that he put his watch on by ten minutes and fined his workers each one penny for being late, thus getting ten minutes extra a day out of them. He was not well liked, but he held sway over the village and villagers alike. Despite this he was a benefactor of the Congregational Chapel at Brookfield, where he had his mill, and was a strong chapelgoer.

He had an only daughter and in order to carry on the family business and name after her marriage to a Major Cuthbert, she took the name of Shepley-Cuthbert. Cuthbert was of good breeding, but not well off. By contrast, her family, although a local family of long-standing mill owners, were more what we now call nouveau riche. The Major was not really interested in the business, more in living the life of a country gentleman. He was passionate about the game of cricket and used to pay the village lads to play cricket on what was the tennis court, below the rear terrace of Gamesley House.

Naturally enough, when old Mr Shepley died, and the going got rough in the cotton industry, they were not able to survive.

The mill is now long gone and a garage and caravan sales area fills the site.

Major Cuthbert had a son and daughter. The son went to America to seek his fortune and the daughter, who never married, stayed to look after her mother, together with a maidservant, following the death in action of Major Cuthbert in the First World War. But money was tight. Gone were the days of the coachman, the two gardeners, the two live-in maids and the two outside maids. That was why the house had not been painted for twenty-nine years. Luckily, the windows were hardwood and were still in fair condition.

Finally, the mother died. The daughter sold up and moved out to a smaller property some distance away. The new family who bought the property at auction did not really have the money to live in the grand style to which they aspired. He was a company director and both he and his wife loved the idea of good living, entertaining and generally having a high old time. It was not long before they were in financial trouble, which is where we came in.

After they had gone, all sorts of people turned up asking their whereabouts, and the threatening post we got was never-ending. In fact, we opened one by mistake to find a request for payment from a Manchester clothes shop, amounting to over eight hundred pounds; this in the early 1960s when an average week's wage was in the region of forty pounds. Even with the gall you require in business, I could never have lived like that. Big debts are not my forte, although some would say that the cost of Gamesley House was one hell of a debt.

Chapter Nine

As I was saying, the villagers would not venture into the grounds, and as all our stock was at the back, they could not even see it to entice them in. And with the high wall, hedge and trees obscuring the front of the house, the passing trade did not exactly materialise as we had hoped. We had advertised in the local papers, but this only produced a trickle. Were we doomed before we started?

Taking a tip from shops that display their wares on the shop front, we decided that, if they would not come to us, we must go to them. So, every Saturday and Sunday morning, we trundled the best of our stock up to the front gates and set it out on tables. Each Saturday and Sunday evening we trundled it all back again.

This was hard graft on top of all the other work required to grow, water and weed our stock, but it paid off. Passing trade would drop in and the villagers came across to see what was what. Obviously, all our stock was not at the front, so, bit by bit, people started to go round the back to the Garden Centre proper. It was not long before lugging the stock round became unnecessary. Trade was definitely picking up. We had priced our stock correctly and although I say it myself, it was good material. The good name of Gamesley Nursery began to spread.

As the months progressed, so did Judy. By midsummer, she had blossomed forth like all the plants we were cultivating and looked clearly pregnant. Still, she worked on pricking out seedlings during the night and selling stock by day. I don't know where she got the energy from.

Meanwhile, Judy's parents moved to Rhos-on-Sea, North Wales, and took a flat overlooking the sea front. This would prove useful in future, meaning we could go to the seaside for short breaks, which we were not likely to be able to afford otherwise. Holidays were not even on the agenda yet, as all our time would be needed to get the business going. However, the occasional break was needed, and those trips to Rhos-on-Sea were a godsend.

On the first occasion we went, we decided to treat ourselves to a meal out. Looking over the various establishments in the area, we saw one a few miles away that we heard had a good reputation, so we booked for the next night.

Arriving at the hotel, which I seem to remember was called The Abbey Hotel, we sat ourselves down to have, a small drink beforehand. Around the walls there were pictures and on examination, we discovered them to be by Anigoni, who had recently been in the news for painting the Queen's portrait. They were for sale and we fell in love with a portrait of a young woman. It was for sale at fifty pounds and Judy urged me to buy it. Now fifty pounds was a lot of money in those days, but with a name like Anigoni on it, we felt it could be a good buy, even though it would stretch me to my limits. So, approaching the barman, I enquired after the picture, only to have my hopes dashed when I was told it was only a print, albeit a limited print. That was too much for us; fifty pounds for a print was not on.

The meal was not too bad, although by the time we got it I was well under the influence, having consumed six sherries whilst waiting. The place went down in our estimation on two counts; damaging our aspirations to own an Anigoni, and making us wait for the meal, which dampened our appetites (but evidently not my thirst!)

On the way back to Rhos I was lolling drunkenly on the back seat when we passed a small dog. Intoxicated as I was, I slurred, 'Hello, dogie-dogie', which caused much amusement to Judy and Cliff. They still rib me about it to this day, whenever I am a little squiffy. Needless to say, the dog didn't hang around for further insults.

Chapter Ten

Curled up in bed one night in September, my dreams were shattered by Judy pummelling me in the back. Her waters had broken. Why does it always seem to happen in the middle of the night? Panic? I did!

Judy calmly told me to phone the maternity home and get together a small suitcase. This part was easy, because she had already prepared the things that were to go in it. We set off for the home and on arrival, got her settled in.

In those days, fathers were not even allowed, never mind encouraged, to stay around for the birth, much to my delight, because I am a coward about these things. I was packed off home by the Matron, with the promise that she would phone when anything happened.

Cliff and I brewed the inevitable cup of tea and with great trepidation, sat down to wait on events. At about six o'clock in the morning, just as we were running out of steam, the phone rang and Matron announced that it was a boy. I didn't even ask his weight.

Leaving Cliff to hold the fort, I dashed off to the home, to be greeted by a smiling Judy with a small bundle sporting a dash of red hair. It must have been the milkman! Judy was blonde and I was brown haired, but I remembered, in time, that my grandfather and my uncle on my mother's side were redheads and Judy also had red hair on her side. So all was well, it was mine after all.

Luckily, I had saved up my holidays for this occasion and took a fortnight off to run the house. I say run the house, it was more like tickling it. I was not the best housekeeper in the world. The washing machine's noises really worried me. I was expecting it to blow up at any moment.

Naturally, we had the inevitable booze-up at the pub next door, after running round all and sundry to tell them the good news. We called him Nigel Peter. Cliff thought Nigel was a naff name, a real women's magazine name, but we liked it.

Despite all this excitement, other things had to be attended to. Around this time of year, autumn, you took delivery of stock for sale in the New Year; trees, shrubs, roses and other plants that you had not grown yourself. Luckily, the growers of this stock, as a custom, did not require payment until the March or April of the following year. This was perhaps just as well. Without this breathing space, I doubt whether we would have been able to continue and this tale would have come to an abrupt end.

At this time we tried another venture. Once you had planted the trees and shrubs in the ground, lifting them could only be done over a short period of time in spring and autumn. The ubiquitous black plastic containers seen today were not even thought of then. We attempted to solve this by taking large, empty fruit-filling tins, obtained from a local bakery, planting the shrub or small tree in compost and setting them up on an ash bed. This allowed the roots to grow through the bottom, but it was easy to lift them out of the ash. The downside was that the tin had to rust away before the shrub or tree could be entirely free. We discussed this idea with our wholesale supplier in Stoke and due to the size of his operation, he was able to order plastic containers in bulk. By the next year one third of his stock was containered and the following year more than half.

We are not saying that we were the originators of containerisation or even that one person was responsible, it seemed to be an innovation, which evolved over a period.

Where would we be today without those black plastic containers? They certainly are a boon to nurserymen, extending the sale and planting of shrubs over virtually the whole year.

Chapter Eleven

Early in 1965, we discovered one of Judy's traits. Arriving home, we were confronted by the sight of her wielding an enormous pair of pipe wrenches, with which she was attacking one of the large hospital radiators, having first already removed an ornamental timber and metal grill surrounding it.

Naturally, we wanted to know what on earth she was doing. She had decided that they needed to go, as we did not and were not likely to use them. This of course became the next project and included the removal of all the pipework as well. We did finally find someone to remove it as scrap, but not with the hoped-for reward in cash.

This was our introduction to Judy's modus operandi and from then on we were not quite sure what we were going to be met with when we got home. Such action was and still is the augury of Judy's intention to decorate. We would find the room in question part-stripped of its paper. The need to decorate then becomes a fait accompli.

However, this did have its moments. Once we had to rescue her when she became trapped between the concrete planks that formed part of a large bench in the conservatory, and which Judy had decided was in the way. The five-foot waif had attacked it with a fourteen-pound sledgehammer, whilst standing on top of the bench. It promptly disintegrated, depositing Judy between two of the planks and firmly trapping her. It was a good job that we arrived home soon after.

1965 was also the year we really started to build on the customers and sales of the year before. Word of mouth was the main way, but some judicious advertising was also used to bring in the punters. We had a bigger stock and the amount we grew ourselves was also expanding.

There were still hiccups. Early in the year we had an enormous downpour. The ground was saturated and we were held up from working outside, but the work inside continued. There is always work to do on a nursery.

Other parts of Glossop were not so lucky. The roads were awash, the bottom main road was flooded. A stream in the woods at the back of Gamesley House got blocked and the pub at the bottom of our hill had water coming in the back door and pouring out the front, together with tables, chairs and barrels. What a mess! Hordes of council workmen, police and volunteers brushed and shovelled the debris into bags and containers to be taken away. I don't know if any barrels went missing, but I wouldn't be very surprised. Life went on, though, and things were soon back to normal.

We approached that spring of 1965 with the usual gallop; seedlings to get out and bedding plants to prepare. We had built a series of frames on our bottom lawn, made from old sleepers and second-hand frame lights, into which we laid boxes of bedding plants. These needed constant attention, lifting the lights off in the morning, watering and putting the lights back on at night or when frost was due. It was hard work.

Easter came, business boomed and we were run off our feet. We even had to send out to our suppliers for an emergency supply of bedding plants.

That summer was good to us too. It rained just enough and the sun shone almost every weekend, our main selling time. We were doing well and we approached the autumn in mellow mood.

Also in 1965, we started on another venture: egg production.

You remember the billiard room over one of the garages? Well, we decided it would make a very good battery hen house. We had looked at free-range, but we would have had the expense of fencing in. Then we looked at hens on litter, but again this was expensive, and would have rotted the wooden floor, so we plumped for batteries or cages. This might have seemed even more expensive, but as this method was now some ten to fifteen years old, second-hand cages were coming onto the market at reasonable prices.

First we had to deal with the billiard table. What to do with it? We decided to move it into the house, in the large, front dining room, which we did not use.

Oddly enough, it came to pieces quite easily, but we had problems with the bed, which came in three sections of two inch thick, heavy slate, not the kind of thing that Cliff and I were equipped to deal with.

Discussing it at work, at Ashton-under-Lyne, where I now worked, a number of my colleagues volunteered to help, and with the aid of a trolley from the Market Hall, we finally deposited the slate bed in the dining room. At this stage, nothing further happened to the billiard table. We were too busy doing other things and there was the question of the hen cages, so we left it in abeyance until such time as we could set it up. That is, if we ever got the time.

We came across an advert for hen cages in the backwoods of Disley, south of Stockport. We arranged a meeting and arrived to find a more or less wrecked hen house, with a set of cages, half of which looked unusable. After much haggling, we bought the lot for ten pounds, quite a bargain, we thought. However, the next thing was to get them back home, which was rather more difficult than it looked. A lot of the bolts were rusted up, and it took loads of WD40 and many bloodied knuckles, together with a set of strong metal cutters, before we were in a position to move them. We borrowed a roof rack and piled them eight sets long and four sets high on top of the Mini.

We set off for home, wobbling and teetering. Cornering was a nightmare. We were sure we would fall over, such were the impossible angles we reached. Don't forget this was not just one trip, but one of several, each managing to outdo the former. Luckily, the police were nowhere to be seen on the numerous trips we had to make, otherwise I am sure we would have been stopped for having a dangerous load. The cost of petrol and the time taken would certainly have made the cages quite expensive if we had bothered to cost it, and we had still not set them up yet.

We erected them in one half of the room, with the intention of increasing the numbers at a later date. We used the best of the available pieces, but still had to discard over a third. This gave us one hundred and eighteen cages. We bought a hundred point of lay hens and installed them in their new home. Having already purchased hen mash from a nearby feed supplier and with water poured into plastic rainwater gutters down each side of the cages from a watering can, they soon settled down.

We learnt one thing about hens. In order to make them lay better you fed them with a small amount of crushed shells. You know, the oyster shell type. This helped them to produce strong shells, although, from time to time we did get some eggs without shells and these, together with the inevitable broken ones, were reserved for our table. We had many a fine omelette.

Chapter Twelve

I mentioned earlier that, in hindsight, we should have bought the two cottages next door. The circumstances whereby this became apparent really is a sad tale.

One cottage was still lit by gas and the old lady who lived there was not very nimble on her legs. Unfortunately, when crossing the road one evening, she was knocked down and taken to hospital. She was in a bad shape and eventually died. This left the cottage empty, and it was put up for sale for renovation and went for seven hundred and fifty pounds.

Shortly after this, the tenant in the other cottage decided to buy, paying nine hundred pounds for the privilege. This made a total sale of one thousand six hundred and fifty pounds, less than two years since we were offered both for two hundred and fifty pounds. We could have done with that money at this time, but it wasn't to be. Ah, well! You can't win every time. Although our property was now worth two and a half times what we paid for it.

Incidentally, at the time of writing, one of those cottages is on the market for forty-two thousand pounds and the main house has just sold for a quarter of a million. At a time when prices are falling, it just staggers you to think that this escalation could happen.

These days there is no way we would even have contemplated carrying out such a venture. It is a shame that this sort of amount never came our way when we sold it. One can always dream!

1965 was also the year when another portion of history bit the dust. Around the back of our property was a railway branch line. It started from the goods yard at the bottom of the side road and served several dyeing and associated works in the nearby village of Hadfield. This year saw the closure of the line and we watched the last train go down and back.

This also gave us an interesting diversion. A group of men appeared with tools and lorries and removed the rails and sleepers completely. It was some weeks later that we received a visit from the police, making enquiries. It seemed that the workmen were thieves and when the proper contractors hired by the railway arrived, there it was, gone. Unfortunately, we had not taken much notice, believing the original men to have been the true contractors. It seemed that everyone else in the village had believed the same. It shows that you can't trust anyone. These days it would be taken for granted that someone would try something and everyone would have been alert to suspicious goings on.

But, back to the billiard table. We decided to set it up as the dark nights closed in, as we felt we might find the time to use it. Some bright spark then questioned whether the floor would stand the weight and so I was dispatched to the nether regions of the cellar and from there, as the cellar did not cover the whole of the house, to the under part of the dining room. All was well, two substantial walls crossed the area and the floor joists alone would have held up the weight.

We set up the base and legs, using a spirit level in all directions, and the time honoured way of raising the legs by the use of beer mats! Then we added the slate bed, but we were loath to use the hard, dried-out cushions. What could we do now?

Here we had more luck. The fire brigade at Ashton-under-Lyne were moving premises and had been provided with a new table, so their old one was redundant. In those days, you never saw full-size tables in the home. The firemen had had no takers, and were open to offers, so, tongue in cheek, I offered them a fiver. Much to my amazement, it was accepted. So I removed the cushions and fixed them to my own table.

You remember that I now worked at Ashton-under-Lyne in the Engineers Department. Ashton is sited half way between Glossop and Manchester. In view of the hard work put in on the billiard table by the lads from my office, we decided to give them a night out. We had the new table, darts, dominoes and cards, plenty of beer and a buffet supper.

The night was a cracking success. It was the night for Morecambe and Wise on TV and everyone decided that they did not want to miss it. So twenty-odd people crowded into the sitting room, sitting or standing where they could. You could not have had a more appreciative audience as we had colour television, one of the first in the village. We were not trying to show off but, as we did not go out, we decided it was the least we could do to brighten up our in-home entertainment.

Later on we realised that, because of the fireplace in the centre of the inner wall, we had moved the table off-centre and we could not play a shot from the window side of the table with a standard cue. We tried a short cue for a while, but even this was not altogether successful. Then someone came up with the answer. When you needed to play a shot on that side, you opened the window, played the shot, then closed the window again; simple. Mind you, people seeing it from outside, must have wondered what was going on.

Darts, dominoes and cards were also being played, the latter two in the lounge.

Later, Judy appeared with supper all piled up on a trolley. There was just one thing wrong. She had opened the top out, but had not swivelled it sideways to support. The hinges were the only things holding it up. The inevitable happened, it broke and a large plate of meat pies and a jar of pickles went floating all over the carpet.

There was silence for a moment. Then someone sniggered and before you knew, everyone, including Judy, was bent double with laughter. They had all had a lot to drink, and were not bothered about a bit of fluff, so they picked the meat pies and onions up and proceeded to scoff the lot. Cliff, the Health Inspector, also ignored the incident.

A good night was had by all and everyone departed in high spirits around two in the morning. Luckily, we didn't get any complaints from the neighbours.

Chapter Thirteen

Entering 1966, we realised that we had a good market for our eggs, so we now looked to expand. Again, we looked for adverts and spotted one for an auction at a farm some three miles away; 'everything to go'.

We had never been to an auction in our lives and were very wary of the procedures. We didn't want to buy something by mistake, so we turned up early to watch. The hen cages were well down the list.

I have never seen such rubbish sell for so much before. It seemed that people got auction fever and bid for things even when they did not want them. An old wooden wheelbarrow with holes in went for ten pounds. We kept mum, with hands in pockets and did not even dare raise an eyebrow. At these prices, things did not bode well for our prize. In order to make certain we were in the right place at the right time, we retreated to the hen houses some four or five lots before and waited with not one other person in sight.

Luck must have stepped in. I don't know what happened, but the auctioneer must have got up a gallop and arrived at the lot we were interested in, with only about six people in tow.

'Where is everyone', he said.

No one knew.

'Well, we must get on. Who will start the bidding at a hundred pounds?'

Not a sound.

'Fifty pounds, then?'

Silence.

'Twenty-five pounds? All right, then, who will give me a bid?'

No one moved.

I looked round, they all looked uninterested, so, in a timorous voice, I said 'five pounds.'

The auctioneer looked at me in disbelief and said 'Did you say five pounds?'

The spectators guffawed, and so in a louder voice, I said 'Yes, five pounds.'

The auctioneer, glancing round, said, 'Come on gentlemen, you can do better than that.'

I held my breath. No one spoke.

Looking over the top of his glasses, the auctioneer, whilst willing them to bid higher, intoned, 'Going once... twice... gone to the gentleman at the front there.'

Off he went at high speed to the next lot, with his small attendant crowd, leaving us in total silence. I looked at Cliff and he at me and we both whooped with delight.

Just at that moment, about five men dashed in. 'Has he got to this lot yet?'

We had to inform them that not only had he been, but he had gone. 'Who bought them and how much did they go for?'

'We did', I said, 'and they went for a fiver.'

It was their turn to look astonished. Shaking their heads, they disappeared in the general direction of the auctioneer and his little band.

The next priority was to find the auctioneer's office and register our purchase, using the little cloakroom ticket we had been given, and to pay up. We finally uncovered it in an outbuilding. Naturally, there was no sign, but judicious enquiries produced the location of the said establishment.

On paying, we were cautioned that we would need to remove the cages within two days or they were forfeit. Bang would go our huge investment, all five pounds of it. Luckily, this time we had made arrangements in case we made a purchase and when we got home we rang a friend who had a lorry to transport the cages back to their new home.

This was easier said than done. There were some three hundred cages and once again the WD40, spanners and metal cutters were pressed into service. It still took us the better part of a full day to get them set up. When the dust settled, however, we had two hundred and fifty-two cages.

We then purchased another one hundred and fifty point of lay pullets. This meant that we could rotate the lives of the hens, so that we would always have some laying.

Food now became a problem. I had been buying five hundredweight at a time, because that was all the poor old Mini van would carry. This was proving to be costly and also, because of the increase in the number of hens, I was making many more trips. The Mini van was not really up to this and repairs were getting more numerous. We decided to take the plunge and buy a bigger vehicle, which would also help in the transport of our nursery stock.

Thinking along the lines of farmers, the obvious vehicle was a Land Rover. We plumped for a diesel, as it was the most economical, and ended up with a five-year-old, ex-farmer's, long wheelbase hardtop. This could carry a ton of food in one trip, making the cost much cheaper. Mind you, as each bag of food weighed half a hundredweight, loading and offloading was a chore in itself.

Unfortunately, the vehicle was not all it seemed, as I soon found out, when, travelling to work, I hit a pothole. Now you would think that a Land Rover was built to cope with just such road conditions but the whole steering wheel shook and it took all my power to hold it, let alone steer. It only stopped when I brought the vehicle to a halt. I proceeded to drive on a little more cautiously, but nothing further happened. I had a look underneath, but could see nothing wrong. So I assumed it was just an isolated incident.

A few days later, I was going down our road across the level crossing, where the former branch line ran, when it happened again.

We took the Land Rover back to the garage and explained the symptoms. They seemed to understand and said this was common in old vehicles and that they would put it right.

When we picked it up, they claimed it was as good as new and we should have no more trouble. Two weeks later, the same shaking happened again, if anything even worse than before, snatching the steering wheel completely out of my hands. Fortunately, I was able to stop before an accident occurred.

It seemed worthless to take it back to the garage, as we had lost faith in them. Being members of the RAC, we asked them for an independent report. This said the vehicle was completely unroadworthy and the garage had broken the law in selling it to us in that condition.

Armed with this, we went back to the garage. They had to admit the faults itemised, but would not give us our money back. They offered instead to sell us another vehicle. We were now extremely cautious of second-hand Land Rovers, especially ones from that garage. We saw an almost new Volkswagen van, which I had heard good reports about. It was dearer than the Land Rover, but faced with the likelihood of being taken to court, the garage did a good deal with us and we went away with our fingers crossed.

We needn't have worried as, except for the usual servicing, plus a couple of small repairs, it lasted us for the next ten years. It was a trusted friend, a good workhorse and only once let us down in a most unpredictable fashion. But more of that later.

Hen keeping created a number of problems, which we had to resolve. The biggest was, to put it politely, manure.

Naturally, hen manure was great for the garden, but you could not just put it on the plants immediately, as it would burn them. We therefore had to create an area to store it until it had broken down and become usable. However, before we could take it out to the store area, we had to clear it from the cage bottoms. This was done by drawing a scraper from one end to the other and collecting the manure in a wheelbarrow. As you will recollect, the former billiard room was over a garage. This meant lugging the wheelbarrow downstairs: not a pleasant task. On top of this, the scrapper for the first lot of cages was continually breaking down.

When we got the new cages, it was decided that we needed a simpler method for mucking out. I set up some wooden rollers with a continuous polythene sheet band. There was a nut on each end of the rollers to attach a handle. We then ran a sheet of tarred paper over the polythene, on which we collected the muck. This enabled us to carry the waste to a low door above the entrance and drop it into the wheelbarrow below.

Mucking out used to take place in the evenings, after tea. As the year progressed, the nights drew in, and we were forced to do the job by torchlight. Also, as time went on, the heap became higher, necessitating the use of a plank to take the wheelbarrow to the discharge position.

On one of these occasions, it was pouring with rain, dark and miserable. Cliff was wheeling the barrow up the plank, whilst Judy aimed the torch. As he reached a crucial point of the plank, Judy heard our dog, Sheba, snuffling and making mewing noises. Swinging the torch round to see what was going on, she discovered that the dog had found a hedgehog and the prickles were not to her liking.

Cliff, meanwhile, was plunged into darkness and totally missed his way. The wheelbarrow careered off the plank with Cliff hanging on to it. He was dragged sideways and fell face-forward into the slimy mess.

Judy swung the torch back just in time to see him floundering out, coughing, spitting and covered from head to foot in oozing hen fertiliser. He cursed and he swore, but soon saw the funny side of it. We hosed him down and he went and had a hot bath. He still smelt when he came down and even the dogs wouldn't go near him. Come to think about it, he might even have a trace of it today. (Sorry, Cliff. Only joking.)

Judy was very careful with the torch from then on. We had a ready market for the eggs and those we did not sell went to the Egg Marketing Board at a preset price. Although this was not very generous, it did cover our costs and the profit we made went to pay for the Volkswagen van. Only twenty-two thousand eggs and it would be ours!

Chapter Fourteen

Towards the end of the year, we decided that a number of trees were in the way, as we needed to extend the frame lights for the next season's bedding.

Our method was to tie a rope as high up the tree as we could, dig down in a circle and cut, hack or saw off the roots as we came to them. Then, when we had attacked enough roots to make it unstable, we would rock the tree with the rope until it came over.

This was very successful with all but one of the trees. This was a sixty foot high yew, which must have had roots going down to China. No matter how hard we pulled, it would not budge.

Then we had a bright idea. The tree was situated at the top of the sloping bottom lawn, next to a path, three feet below the top lawn, which itself was five feet below the back terrace. We planned to attach the rope to the back of the Volkswagen and rock the tree back and forth with the vehicle to loosen it.

We extended the rope and fastened it to the towing bar at the back of the Volkswagen, which was positioned on the drive at the side of the house. Cliff took up position halfway along, armed with an axe ready to cut the rope if the tree decided to go. Judy was down by the tree to watch its progress and to shout a warning if necessary. I manned the van, started the engine and using the clutch, proceeded to rock the tree back and forth.

Judging that we must have had some effect, we unhitched the rope and tried once again with manual power. Nothing. It would not even tremble. So, back to van power. We all took up our respective posts, but just as I was about to start the engine, some friends of Cliff's arrived.

Leaving the van with its brakes on, I got out and shouted to Cliff, who climbed up onto the drive to greet them.

In the meantime, Judy heard a creaking noise and realised the tree was moving.

'It's going,' she said. Unfortunately, she said it in her normal voice and by the time she had repeated it in a shout so that we could hear, it was too late.

The tree keeled gracefully over, heaving on the rope, which slid the van backwards, even against the brakes.

We turned round in time to see the van describe a graceful arc, for all the world like an Olympic diver, over the banking and onto the lawn. It landed balanced on its back end with the nose pointing into the air, just as if it was ready to take off like a rocket to the moon.

The tree was down, but now we had to get the van back up onto its feet. Even if we could, was it damaged?

We couldn't do anything, we hadn't the tackle. So we rang the local garage, an RAC agent, and asked if he could tow us off, as our van was stuck on the lawn. Of course he would, no problem. But he hadn't seen it yet.

When he arrived, he drove up the drive to the back and staring in astonishment nearly joined our van down the banking. He just stopped in time.

'How the bloody hell have you managed that?' he yelled.

He attached a chain and delicately pulled his vehicle forward. The van gently leaned over and settled with the grace of a ballet dancer onto the grass bank. Taking the strain, he pulled the van the rest of the way up onto the driveway.

We inspected the damage. Pulling off the tufts of grass, we discovered no dints, not even a scratch. But what about the engine?

In Volkswagens they are in the back, exactly the part it had fallen on. Checking that the oil level was all right, with crossed fingers I tried the starter. It fired first time. The gears were also fine and it drove around without a wobble. Now for the bill, we thought.

'Don't worry about that, you are in the RAC. How about I say I pulled you out of a ditch?' he said, trying to keep a straight face. So that was that, our prayers were answered again.

Chapter Fifteen

In the autumn, we received a shock. While serving customers, I approached a man and his wife and asked if I could be of assistance.

'No,' he said, 'I'm looking you over. I am going to start a new garden centre up the road in the next village.' The bloody cheek!

He proceeded to tell me he would soon have us out of business as ours was only a small concern and his would be run on much more professional lines. He was also into turf and laying turf, operating from a place near Manchester.

That evening, we were down in the dumps. After all our careful planning and just when things were getting better, this happened. We plotted sabotage.

As usual, Judy was the one to break the gloom.

'We are not going to be beaten at this stage,' she said. 'He can't start till next year and it takes at least a year to get settled into operations. We have got to look at ways to make ourselves more attractive and accessible.'

With our business marketing heads on, we decided that we needed to open up the front, make better paths and create a road down to the main garden. We had planned on this before, but we were waiting until we could afford it. Now we were being forced to do it, if we wanted to stay in business.

We had never had a bank account up to that time, paying our way by cash. Now we needed one and one which would lend us some money, quick!

There were the big four banks in Glossop. We made an appointment to see the manager at each. Three turned us down flat, even though we personally banked at one of them. The fourth was prepared to help, as long as Cliff and I would change banks to them.

'What do you need?' they asked.

We had no idea. Naive as we were about business loans, we

had not yet calculated the cost of getting the work done.

The manager proposed we got a price and came back with details and suggested a contractor who might tender for the work: one of the bank's customers, obviously. They might as well get it both ways!

We were familiar with the man, he was the son of a local farmer, but were put off by tales of his behaviour. He was a hard drinker and known to have been in trouble with the boys in blue on a number of occasions. But we knew of no one else, so we phoned him.

'Yeh, no problem,' he said. He would come and see us.

He arrived two days later looking like a reject from a hippy commune. It did not inspire us with much confidence in him. We explained what we wanted.

We needed to clear the trees from the front, including the privet hedge, drop the height of the front wall, form a path up the middle and pave the front terrace, re-turf the two sides and clear a clump of trees at the rear of the house, and create a road down to the back gardens. Just a little job, really!

We warned him that the electricity supply was overhead and passed through the trees at the front, and likewise the telephone wire. Also there was a flagpole, which we were desirous to keep intact.

Armed with this information, he went away and promised to come back within the week with a price. As it was, he did not appear with the price until the following week. This bode badly for the progress of the work.

Supplied with his price, which to be fair was very reasonable, we returned to see the bank manager. We also wanted money to cover for additional stock, which we hoped would help us pay off the loan quickly. He seemed very amenable to our requirements, but pointed out that it was not up to him to agree the loan.

'Of course, its head office's decision,' he said, abdicating any responsibility for an adverse reply.

He telephoned us three days later and asked us to meet him. From his tone, it did not sound like good news. He explained that he had tried very hard to get what we wanted, but they had overruled him and would only loan us the cost of the works.

This seemed very petty. As part of the arrangement for the loan, we had had to pay for a valuation on the property, which would be used as collateral. This showed that, after taking out the mortgage, the value of the house was, at that time, at least three times what we wanted to borrow.

But we were faced with another fait accompli, the other banks would not even look at a loan. In fact they would not even talk to us (maybe it was that hen manure). So we had to agree to whatever loan we could get and get on with the job.

The loan was to be paid back over a five-year period and projecting our profits and assuming that nothing went wrong, we could just about do it. What could go wrong? We should have considered our recent past. Things were not auspicious.

We phoned the contractor and informed him that we were now ready to proceed. Could he give us a date? Just like any farmer or contractor he was very noncommittal, only saying that he would start in the spring.

I wondered what he meant by spring? Probably midsummer if my experience with builders was anything to go by. We would just have to wait and see.

Chapter Sixteen

Christmas 1966 was now upon us. As usual, both sets of parents were invited to our house for Christmas day.

At two-and-a-bit years old, Nigel was beginning to understand what was going on. We were not aware how advanced he was until we set up our Christmas dinner on our biggest table in the lounge.

We were all sitting around having consumed enough Christmas fare to feed the whole village; the cake, mince pies, cheese and biscuits. Nuts and celery in a glass vase were scattered amongst the table debris, but everyone seemed too full to eat more.

Like most mothers at Christmas, Judy urged us to have some more, so Granddad Binns reached out and took a stick of celery. He put salt on his plate, dipped the celery in the salt and proceeded to take a large bite.

Nigel's face was a study of horror. After a pause, he blurted out, 'Granddad's eating the flowers,' with obvious indignation.

From an early age, we had instilled in him that he should not touch, let alone eat, the flowers and here was a senior member of the family doing just that. It took us a long time to explain that these were not flowers. With a nursery garden full of plants, we could not afford any damage. He was still not entirely convinced, but we would have to watch him in future, to make certain he did not follow Granddad's example on some of the more poisonous plants.

It was at Christmas that we learnt that the contractor would start in early spring, when the weather was right. In the meantime, we had approached the County Agricultural Division to obtain permission to remove the trees, as some of them were protected. They had agreed, as we were planting a number of smaller trees on the front, to replace those lost and we would have far more trees in the grounds for sale.

It occurs to me now that we never did apply for planning permission for change of use to a nursery garden centre, let alone apply for that sign on the front. Perhaps the Council were grateful for a business in the town, especially one that looked to the future of an expanding population. I doubt that they would be so magnanimous in today's climate.

So it was that we moved into the New Year with a degree of optimism. One bit of news was not really welcome though. Old Stanley, the ex-farmer, who had worked part-time in the Garden Centre for the last two years, decided he was a bit too old to carry on and so at the age of seventy-six, he retired. We were going to be hard pressed to find a replacement.

Chapter Seventeen

1967 looked like the year we would have to pull out all the stops if we were to remain in business.

The contractor actually turned up on time and work started in the early part of the season on the clearance. The wall was partly taken down and the stones deposited on the lawn. Then the privet hedge was ripped out and carried away. Next he had to tackle the trees.

Once again we reminded the contractor about the electric and telephone cables, not to mention the flagpole. He seemed disinclined to call any of the utilities in, claiming there was no problem, he could take the trees out without disturbing anything. We remained sceptical. Naturally, with all the work going on, the local populace were out in force to see the spectacle. They were not disappointed.

He was taking out a tree in the middle of the front garden when suddenly it went the wrong way. Crashing down, it cannoned into the flagpole, which promptly split in half, sending the top part sailing through the air straight through one of the top windows. This not only broke all four panes, but the centre bar to the sash window as well. Horror! It did raise a cheer from the onlookers, but a definite groan from us.

'Never mind,' he said, 'I'm insured. I will have it put right.'

'You better had,' we said.

Luckily, this was all he could do for the day, so we repaired back indoors to lick our wounds.

The next day dawned bright, but extremely breezy. We thought, He can't do anything today, it's too windy. But no, there he was complete with tractors, ropes and men.

After the spectacle of the day before, the news had got round the area like wildfire and it seemed the whole village was out to watch today, even bringing chairs out for a grandstand view.

Now he had got round to the trees through which the telephone wire passed.

'Shouldn't you call in the telephone people to remove the wire until you have finished' we said.

'No need, we'll be careful.'

Within ten minutes the first tree was down, complete with phone wire.

There were cheers from the watching crowd. They'd not seen anything like this in years and it was all for free.

'Don't worry, we will have it reconnected,' said the contractor.

By now, we were past caring, so when he got round to the trees that the electric cables passed through, we did not utter a sound, just crossed our fingers and offered up a prayer, to no avail.

Three highly active electric cables promptly flicked across the highway, bringing everything to a halt and eliciting the biggest cheer yet from the onlookers. This definitely was more fun than *Coronation Street*.

We had a powerful three-phase supply into our house. Hissing and spitting, the cables gave quite a good firework display. Nothing could pass and within a very short time we had the police in two cars and a fire engine that someone thought we might need. An emergency call went out to the electricity board and while the traffic continued to pile up, everyone chatted to everyone else about what should be done.

Once again we heard a certain gent say 'Don't worry, I am covered by insurance, we will get it put right.' If his insurance company had been around, I am sure they would have cancelled his cover on the spot.

The electricity people turned up to find a huge crowd of people ready to applaud their endeavours. With some difficulty they managed to switch off the electricity and remove the cables, but they were not prepared to wait around until the rest of the trees were removed to reconnect. They left.

This left us in a quandary with no power. Finally, after many phone calls, which had to be made from a public phone booth, as ours was not functioning now, an inspector turned up.

First he ascertained that the contractor would pay and obtained a signed statement to that effect. Then he approached us.

'Do you really need a three-phase supply?'

We had no idea. As we escorted him round the house, he did an inventory of our appliances. He felt that a single-phase would do, but he would replace the three-phase if we wanted. We just wanted an electricity supply, full stop; single, three or fifty-phase, it did not matter. So it was agreed that he would arrange for a single-phase supply. But when? Radioing back to base, he found out that the work could not be started until the next afternoon and would possibly not be finished until the day after.

So there we were; two nights without electricity. What was the contractor going to do about that? Like all workmen, he just shrugged his shoulders.

We went across the road to our local shop, where we bought in a full box of candles. Mind you, these were a good investment as the number of times that the overhead electricity cables cut out in the village was legendary.

Now the contractor had removed the wall, hedge and trees, one of the villagers commented, 'I'll be able to see into your house a lot better now.' The cheek!

What more mayhem could our erstwhile contractor achieve? It did not take long to find out.

Having dispensed with the trees at the front, he needed to start on those at the back. There was no road down to the back garden. In order to obtain access, he had to carve a road down by the side of the back lawn, which was supported round the sides by a stone wall. Sending in his JCB, the operator steadily hacked out lumps of the garden until he knocked the wall down. We should have guessed. We did not even ask him what he was going to do about that, the answer would have been 'Don't worry, I will fix it.' Fortunately, we did need the retaining wall rebuilding, but we were not going to tell him that. We decided the sooner he finished the better, as our state of mind was being seriously eroded.

The trees at the back were easily dispensed with. Luckily nothing more went wrong. It took some four months and some serious pestering before he came back to lay the turf at the front. We reluctantly paid up and thought, Never again!

Now came the big question. Would it be enough to attract the customers? Time alone would tell.

Chapter Eighteen

During the year, much to Judy's delight, we finally got round to doing some decorating and renovation.

I have not really told you much about the house, apart from the fact it was large, too large really. The main portion had a full-pitched roof with a smaller, flat-roofed, side section. In order to create the illusion of being long and low, the pitched roof stopped short with a large, lead, flat part on top. From the front, this looked like a ridgeline, a clever deception. You could see for miles stood on the top and Cliff wanted to put up a telescope, but we had no time for just looking.

The house had a sitting room, a dining room (now the billiards room) and a study across the front, with three bedrooms of the same size above. There was a side entrance porch leading into a grand central hallway, with a toilet and second staircase at the other end.

At the back as you came into the house was a lounge with a bay, with the main bedroom and balcony over. There was a magnificent staircase with a doorway leading into the old butler's pantry and on into the main kitchen. Two bedrooms above this area had been the live-in servants' quarters many years ago. Finally, from the kitchen you could pass into the scullery and pantry, with a bathroom and separate toilet over.

The house had other innovations, other than the roof, some well ahead of their time. The upstairs water cistern went into an internal sewage pipe in lead, which also took the rainwater from the flat roof and the overspills from the main pitched roof, which emptied onto it. This pipe had a wonderful mahogany surround and the water cistern was a lead-lined mahogany box. The whole thing was set in a carved mahogany enclosure behind the WC pan, which also had a mahogany seat.

The rest of the room was panelled up to the same height as the cistern covering. The effect was reminiscent of the Victorian toilets in Pullman coaches on the railway. The bath itself was a centrally located, six foot six long cast-iron affair, which allowed you to lie in luxury, full-length with ease. It was also great to bathe kids in, as you could approach it from both sides.

Among the other unusual features were the two fireplaces in the front sitting room and the study. In a modern house you would position radiators under windows to allow the air currents to take the heat round the room. In this house the same thing had been done with these two fireplaces, with the flue taken up the side. It was quite odd to sit there and look through a window whilst also looking at the fire. There were the usual blind windows where a bricked-up indentation appears where the window would have been but in this case it was done with stone. This set the elevation off.

The lounge bay had a huge, central, sash window, the sill of which was only inches off the floor. The bottom sash, when pushed up, allowed you to step out onto the rear terrace, a forerunner of modern French and patio windows.

All the ground floor windows had concealed, hinged doors in the side jambs, which could be closed across the window with a bar to lock them in place. This conserved heat in winter and was also a good deterrent to burglars.

The ground floor was twelve feet six inches to the ceiling while upstairs it was ten feet six inches. There was the usual neat coving to all ceilings, but the small sitting room at the front also had an ornate ceiling, apparently made of plaster. It had panels set with a curved design and a rose between each set of panels, like a palatial country house. It was only when we came to decorate this, we found that the panels were formed in papier mâché held up with copper nails. We decided to do the right thing, so we painted it white on a dark green background. I ended up perched up on a plank between two ladders, flat on my back. Now I know what Michelangelo felt like painting the Sistine Chapel in Rome. My neck ached for days. However, it did look fine when we had finished.

There were also cellars under the hall and the rear of the house. These had stone slabs for food, meat hooks in the ceiling (no fridges in those days) and a wonderful wine cellar with a barred door. It just looked like a cell. It was a good frightener for any unruly children. If they misbehaved themselves, we threatened to lock them up there.

We obtained a new kitchen, courtesy of a foreigner. No, not a person from overseas. A 'foreigner' was what we called someone doing a job outside his or her normal job. In best 1960s fashion, the décor was bright orange. We thought it looked good at the time, but to our eyes today it would be hideous.

Whilst we were fitting the kitchen out, we found, hidden behind brickwork, a black lead grate. Unfortunately, in those days, these were not considered to be an asset. We ripped it out, so that we could use the space. Nowadays that would be criminal.

We did not get very far with the decoration, just the sitting room, kitchen, scullery and our bedroom. We just did not have the time, and could not afford proper decorators, so, like the Forth Bridge, we were going to have to do it a bit at a time, until we finished. Then we'd have to start all over again.

The interior of the house was a bit like the outside, in that it had not been decorated for some time. Walls were generally dark colours, greens and browns. Why did the Victorians feel that their houses should be so grim? They painted them with colours that were very dark and then usually had huge trees or bushes outside which made the general atmosphere even darker. We felt that colours should be light and airy, that sunshine should be allowed to get in, but this would take time. In the meantime, we would have to put up with it.

Chapter Nineteen

By late summer, we were finding the loss of our old ex-farmer friend a bit of a bind. We really needed some help.

Luck was with us. A chap who lived in the village had just lost his job. He was a plumber by trade, but had been a jobbing builder, doing all sorts of maintenance work. We already employed his wife, part-time, in the house, as Judy could not do everything and he, being used to working, was not going to just sit around. The building industry was in the doldrums and there were few jobs around, especially in Glossop.

At first we were apologetic, we could not afford a large wage and we were not likely to match what he had previously been getting. He was not worried, just glad to get some kind of job and he lived across the road.

As I have said before, stock was bought in autumn, but not paid for until spring. If only companies were so flexible nowadays. However, we still needed to transport the stock from the growers to our Garden Centre and in the past I had been able to borrow a lorry from a friendly builder. Sadly, he was not able to help us. Our new gardening assistant knew that his old firm had a lorry of some vintage, which he was sure he could borrow. So it was arranged.

Did I say 'of some vintage'? He rumbled into our driveway with this thing. It appeared to be a close relative of a Model 'T' Ford. It smoked, rattled and looked as though it would have been better off on a scrap heap. But who were we to look a gift horse in the mouth, especially as it was free?

Judy set off with our new helper in this monstrosity, as Cliff and I had to go to work. We were not sure whether we would ever see the two of them again. We could not envisage the vehicle making Stoke-on-Trent where the grower was based, never mind getting back.

Both Judy and our erstwhile helper were smokers. They had

no problem with the lack of an ashtray, as there was a damn great hole in the floor through which they were able to flick their ash. Also, there was no danger of a smoky fog building up in the cab because the windows would not shut. By a miracle, they arrived at the wholesale nursery. Having loaded the lorry high with trees, shrubs and roses, it took on a decidedly strange tilt. Judy was sure it would not even be able to pull away.

But start it did, and so did the rain. And the wind. Even the elements were out to make life difficult! Trundling along, with frequent lurches to one side caused by the ancient suspension and the overloading, was not enjoyable, on top of which they were getting wet through those open windows. Suddenly a huge gust of wind ripped the driver's side wiper off. Although emergency stops would have been impossible, Bill (I'll call him that to avoid lawsuits) scrambled out to see if he could find it. Surmising that it had gone into the back of the lorry, he climbed aboard, but, try as he might, he could neither see nor get in past the load to look where it might have gone. There was nothing for it but to continue without the wiper. Unfortunately, there was no passenger side wiper either. What did you expect? At least he would have no difficulty in looking out of the driver's side window. There wasn't one.

The next problem was a near miss with the gentlemen in blue. Luckily, Bill spotted them before they spotted him, and he sped (or tottered more like) down a side turning. They had no idea where they were going, but were well aware that if the police had stopped them, they would have been guilty of overloading, driving without windscreen wipers, faulty this, as well as that, and they would have been going nowhere. Going somewhere, even if they didn't know where, was preferable. They had no map, so they took potluck. After several lanes and farmyards, they arrived back on the main road with no sign of the police.

Coughing and wheezing, the old girl finally made it home much to everyone's amazement and relief. By the time that Cliff and I got back, everything was unloaded and the lorry returned to its owner, who completely dismissed the lack of a wiper, saying he would get round to putting it right sometime.

From the state of the lorry, it looked like he had been saying that for a long time. It was a good job they did not have MOTs in those days.

Chapter Twenty

One benefit of this house was that there was a pub next door. Not that we ever got to use it that often, but it was the only pub in the village. In the late summer of 1967 we heard that the brewery were going to knock it down and build a new one.

Now normally this would not have caused much consternation, after all, there were other pubs not far away. But the villagers had always celebrated everything in that pub and we had enjoyed some grand affairs since we had arrived there. Anyway, this year was different, because this was the year when the law against drunk while driving and the breathalyser were introduced. Now nobody was going to take the chance of driving to another pub. Much discussion took place in the pub and the village as to what to do for the Christmas festivities. There was talk of hiring a bus, but all the other pubs we could go to were also celebrating and villagers being villagers, they did not really want to mix with those 'down t' road'.

Judy suggested that we could hold a party in our house, it was big enough after all. But we could not afford the booze and food. After much scratching of heads, it was proposed that a charge be levied on everyone coming, to offset the cost. This sounded fine, until one bright spark pointed out that we might not be breaking the law on drink driving, but we would be breaking the law by charging, especially as it was for alcohol.

The landlord of the pub came up with a suggestion. He had applied in the past for an occasional licence for a non-licensed venue. Why didn't we apply for one, quoting the special circumstances?

Well, nothing ventured, nothing gained. We applied after talking to the police Superintendent and getting his backing. I am not sure whether we would have been the only house in the country to try and get permission to act as a pub for the night, but I bet there were not many. We got the licence.

An ad hoc committee was set up and plans made. All the drinks were ordered and glasses hired for the night. Music was my department and the lounge was set up for dancing. A bar was organised and beer barrels were racked to settle. A buffet was to be in our sitting room, and of course, we had the snooker table and darts. Woe betide the snooker player who got in the way of the darts, as they were both in the same room, bringing a whole new meaning to being snookered!

Judy was worried. Would it go all right? Had we got everything? Was there enough food and drink? Would anyone actually turn up? Cliff and I were beginning to pray that it was all over and done with. Nothing else seemed to matter. It was a great responsibility for her and us.

The night arrived and so did the villagers – in droves. We must have had at least two hundred people wandering about the place. The lounge floor bounced up and down to the dancing and I was convinced it would cave in. People came from miles around. How they got there or how they got back we don't know.

Luckily we had a good reception committee who vetted who came in and in those days we had not heard of rowdies gate-crashing parties. In this village as in most others, people did not lock their front doors. They even left the door open in nice weather and went out for the day. It is all a far cry from today when you could be attacked even in your own home when your door is locked.

There had never been a do like it, at least that's what the villagers were saying. In fact there was much talk that we should 'do it again the next year', but by then the new pub would be open. We believed this had been a one-off and as such, a licence would not be available in the future.

The villagers came in early the next day, even though it was Christmas Day and helped clear up and tidy up, so we could have some semblance of normality for the rest of the festive season. What a year it had been. What would the New Year bring?

Chapter Twenty-One

1968 dawned, and we realised that we needed to replace one group of our hens.

In the past we had always bought in what are called 'point of lay' birds. This means that, at purchase, they are not chicks but are almost mature hens, ready to lay. Within two to three weeks they start producing eggs.

However, buying this way was a costly business. Ever trying to save money, we decided to have a go at raising them from day-old chicks. So we set up a timber-slatted floor and a large, netted cage in an outhouse adjoining the top garage and purchased two hundred chicks.

At first, we were worried that the chicks would fall through the slats, they looked so small, but somehow, they managed to leap about between the boards, their little wings beating furiously in an attempt not to plummet to the depths. Caring for them was a chore, as we had to feed and water them on a twice daily basis.

It was amazing how quickly they grew. One moment they were small, fluffy, yellow balls and the next they were the same size as our battery hens. We were tempted to move them, but we had to choose our moment and the old hens were still laying well.

Eventually, the time came to get rid of the old hens. This was done by contractors, who took them away, but they only gave us a pittance. Mind you, we could not complain, we had had our money's worth out of them.

We installed the home-grown hens, and they soon settled down. People think that it is cruel to keep hens in batteries. In some cases, farmers could be just that cramming many hens in each cage, but as we put only one hen in each, they had plenty of room to move around. If a hen got out, the first thing they did was to frantically try to get back into the cages. They were well fed and watered and temperature and light was kept as near even as possible. If they were not kept happy, they would soon stop laying.

We came to the conclusion that the work involved in raising further chicks was too much to fit into our already tight schedule. In future we would buy 'point of lay' and to hell with the cost.

We now had to dismantle the chick cage and clean up. Considering those yellow balls of fluff were so small, It was remarkable how much manure had built up over such a short period. It would all have to be barrowed to the heap. We entered the outhouse and started to dismantle the main cage. Suddenly, Judy froze. She had noticed a movement in the muck under the slats.

Looking closer, we realised what it was. The whole area was heaving with enormous rats. Where the hell had they come from? No way were we going to continue the job whilst they were there. We decided that they might not be around after dark, and so we intended to take in powerful torches to frighten them away. A naive attitude you may think, but it worked. There was not a sign of them that evening.

We worked late into the night at what could be called colloquially 'shit shovelling' by torchlight. It was quite an experience. We were tired, but pleased at what we had managed to do and we hoped that the rats would stay away. We never did find out how they got in there in the first place, but we weren't going to try raising chicks again! What we could not understand, was why the chicks and hens had never seemed to be bothered by the rats.

With all livestock, but especially with hens, you do get ones from time to time who perform poorly and these we had to 'deal with'.

The first time, Cliff tried the time-honoured technique of wringing its neck. However, all that happened was that the neck seemed to stretch and stretch, the poor hen's beak wide open, but no broken neck. In despair and to relieve the hen's suffering, Cliff dispatched it by humanely cutting off its head: not very pleasant, and yes, it's true about headless chickens. On the next occasion, we discussed what to do and so I had at go at breaking its neck. The same thing happened.

Judy, in despair, said, 'This is how you do it,' and grabbing the poor hen, just went twist, pull and it was done. Naturally, we designated her chief hen 'putter-downer'. Not, as I say, that this happened very often. She was not very happy about it, but she could do it, we couldn't.

I would have to watch my step in future if she was such a good killing machine!

Chapter Twenty-Two

Had all the work on the Garden Centre paid off? Yes, it had. The public just rolled in, although the good spring weather had helped. Now that they could see into the grounds, passing trade increased, and word-of-mouth advertising did us a power of good. The whole enterprise was really taking off.

We needed more land to cultivate and having cleared a group of trees and a wall, it was now possible to move on to the paddock ground to the side of the property.

This was rather stony ground and had never been worked. Although Cliff and I had grown muscles we had never had before, it was going to be a bit much to work it over by hand. We resolved to go into mechanics and buy a rotavator. We sent off for all sorts of brochures, but most were out of our price range. Some had several gears, knobs and levers. We just wanted something simple; start the engine, let out the clutch and go.

We eventually found it in a Merry Tiller by a firm called Wolsley. The beast had various add-ons that could turn it into a mower or you could even fix on wheels and a trailer. We already had a mower, but we liked the idea of a trailer, so we purchased one from an agricultural firm near Stockport, complete with the trailer option. The rotavator made short work of the hard ground, but kept on pulling great big stones up, which not only stopped the engine, but necessitated humping the stone out of the way.

However, the use of the rotavator did create one quite memorable episode. The year was very warm and working in the garden was heavy and hot work. Judy was wearing a bright orange bikini, complete with a wide-brimmed floppy hat. She decided that she wanted to have a go with the rotavator, so, resplendent in orange bikini and bright green wellies, she waded in and proceeded to do quite a good job. However, the sight of her grunting and puffing with the machine's vibrations shaking parts of her anatomy, was just too much for us and a number of goggle-eyed customers. We all started cheering, much to Judy's chagrin. She abandoned the job and retired blushing, into the house.

As for the trailer, this was fine, apart for one thing. The land fell away from the house down into the grounds proper and the damn thing had no brakes. Many a time I nearly had an accident. One time I misjudged my speed and tried to take a right angle bend. I went round on two (or was it one?) wheels, almost depositing a collection of just potted Rhododendrons. Judy would have been furious, having spent most of the afternoon on the work. But that trailer did have its uses.

We had much need of the mower, as a good fifth of the garden was lawn. On top of this, both the front and back lawns had an embankment of grass, the rear one being some four feet high. Up to this time we had employed an ancient scythe which we had discovered in the old potting shed. Both keeping it sharp and using it was very time consuming and time was not something we had a lot of.

At this time, hover mowers were gaining in popularity and the prices had come down considerably. We bought one and it proved to be a boon. Swinging on the end of a rope, it made short work of the banking.

Other work, which constantly needed doing, was the weeding of the paths. In those days weed killers were rather limited and the time-honoured way was to use Sodium Chlorate. One time Judy had sprinkled the Sodium Chlorate crystals onto the weeds at the edges of the paths in the morning and left them. After lunch she was showing customers round and as usual, she was smoking. Having finished the cigarette, she threw it down on the path. There was a fizzle and a pop like Chinese firecrackers. Sodium Chlorate is flammable and the whole path exploded in a firework display. Judy struggled to put it out, bending down to smother the flames. Judy had long, blonde hair and without warning, this started on fire as well.

There was pandemonium as everyone fought to quench both Judy's hair and the path. They were both finally brought under control. As luck would have it, Judy's hair was only singed. If you ever use Sodium Chlorate, make sure no one sets it on fire by carelessly throwing down a match or cigarette.

Chapter Twenty-Three

At the beginning of summer we lost our Bill. He had at last found full employment with another building firm and as it paid almost twice as much as we did, one could hardly blame him.

However, we were again fortunate. The head of a family who lived in the village had retired from work. He was an enthusiastic gardener and many times he had come across to talk about horticulture. In fact, we had become quite good friends. So when he heard that we needed some help, he was keen to come to work for us. This suited us fine and we were back to strength before we knew where we were.

He was a marvellous character. He was prone to small accidents, like pulling his hankie out and spraying pound notes like confetti around the grounds, or putting tomatoes in his pocket and forgetting about them, and then sitting down on them.

He once broke his glasses and had to tie the arm to the frame with string. He went to the opticians for a new pair, but had to wait, so he went around for days all tied up with string. After some time he went back to enquire about the new ones, but as they were still not ready, the optician offered to give him a new piece of string. Those were the days!

Having a new helper around, we considered a holiday, after all, we had not had one for four years apart from the odd short trip to stay with Judy's parents at Rhos-on-Sea. We could not afford too much and we had the problem of the two dogs. We felt that it would be unfair of us to lump them onto our new employee, so we booked a holiday on a boat on the Thames with Salters at Oxford, having first ascertained that they would allow dogs on their boats.

(Note: we did ask them if they allowed dogs on their boats, we did not tell them that we had dogs in the plural. After their initial shock at seeing two fully grown labradors, they were quite amenable to our having the boat.)

By now we had moved on to a Morris Minor Traveller, as well as having the van. This made travelling to Oxford, loaded with dogs, child and sundry gear, easy. We boarded our craft and having had some instruction, set sail downriver towards Abingdon, where we were to stay the first night.

We actually made it to Windsor during the week, despite breaking down in the middle of nowhere. The damn engine just coughed and died. We found ourselves drifting slowly downstream. Panic! Where was the next weir? Could we stop in time? Luckily, the drift of the river took us into the banking amid a veritable jungle of reeds. We eventually found the bank itself amongst those reeds and moored up. But where were we? Climbing with some difficulty to the top of the bank, we found we were near the posh public school of Radley. I wondered whether we were trespassing, but it could not be helped. It was miles from any road, and was a long walk to find a phone.

The mechanic was none too pleased to have to walk a mile and a half from the nearest road, but he did manage to put things right, so that we could continue to an overnight stop at Abingdon.

We had breakfast on board every day, then progressed downriver. We would go to a hotel or restaurant for lunch, then onwards to our evening stop, where we had a snack for our evening meal. This was fine. Even with our son Nigel there was no difficulty in getting into a pub for the lunch meal and drinks.

However, at Windsor, we moored up on the banks of the grounds of the famous 'Trout Inn'. Hardly had we stopped, than an irate gentleman descended upon us and berated us for daring to moor there. With quiet disdain, we pointed out that we were desirous of partaking of a meal at the inn and were quite within our rights to moor there. What a change! We were ushered into the restaurant and really made welcome. They could not do enough for us, so we had the last laugh.

On the way back, we had a small incident with Nigel. We were moored up at Maidenhead for the night and he was playing around in the cockpit. I had warned him about the dangers of getting too close to the edge but, boys being boys, the inevitable happened and he fell in.

Now, I had made it my business to teach him to swim at an

early age and he was like a fish in water, so I was not really perturbed. I moved to the rear to help him out when he surfaced.

However, the occupants of the boat behind only saw a very small boy fall in. Fearing the worst, the gentleman of the party flung off his jacket and before I could stop him, dived in and pulled Nigel out. Unfortunately, he was wearing glasses and while diving in had lost them on the river bottom.

I thanked him very much, but did not mention that Nigel could swim (it did not seem appropriate). I offered to pay for new glasses, but he would hear nothing of it. He was just glad he had been able to save Nigel's life. If the gentleman reads this, obviously having got new glasses, I thank him once again.

We returned to Oxford and then home, highly refreshed and ready to tackle the hordes of customers.

The Garden Centre was indeed booming, and the worries we were expecting from the rival nursery up the road, never materialised. We were doing very well, thank you.

Other things did happen though. The projected Manchester overspill estate was due to start construction in 1969 and we had an awkward spin-off from this. The farmer who owned the land where the estate was planned had never been particularly kind to his animals. Some even said he was cruel. With the advent of the estate, he was not keen to keep on top of maintenance, and fences were forever falling down. The result was that we, on several occasions, had sheep or cattle straying into our grounds. Ringing him up to remove them did not have any effect and we were forced to round them up ourselves and drive them back to pasture.

On one particular occasion, the ground was wet and the cows had made large holes in the rear lawn. We phoned him and nothing happened. So we phoned again and threatened to impound the animals for compensation. That did it. He was round in two ticks and even apologised. Hit 'em in the pocket I say! It works every time.

We had one laugh, though. Judy heard a peculiar noise coming from the adjacent side road. Now, from our backyard there was a gate onto the side road. This was a solid affair and Judy could not see what was going on. Being of an inquisitive nature (nosy really), she opened the gate and peered out. Seeing nothing, she advanced further, only to fall over a pig, which had been lying in the gateway. In her defence she said, 'You don't expect to fall over a pig at your back gate.' The pig, incidentally, got up with some alacrity and ambled away. And so we came to the end of another eventful year.

Chapter Twenty-Four

In early 1969 we discovered that one of the dogs, Neta, was pregnant. There was a local male golden labrador from the next village whom we had caught a number of times attempting to do the dreaded deed, but we must have missed it on this occasion. Luckily, we suspected that it was this golden labrador who was responsible, so we would not end up with a 'Heinz fifty-seven.'

The pups were due in midsummer, and so we would have to wait to see if we were right. However, we were worried, as both dogs were highly strung to say the least and we were not sure Neta could cope.

It must have been catching because shortly after, Judy announced that we ourselves were due a happy event, again in late autumn. We were much more prepared for this one, although it meant that the new baby would be due about the same time as Nigel's birthday. We did not know if all the fuss that goes along with a new baby would cause him problems when he expected a fuss for his birthday. Neta having pups would be a good way to gently introduce Nigel into the mysteries of birth before Judy got to that stage.

Meanwhile, life had to go on and the Garden Centre was still doing well. This was the year that the new estate started to take shape. It soon became apparent that the type of people who were to be rehoused here were not of, shall we say, the better class of tenants. The open-door policy of the villagers soon became a thing of the past.

However, there were still those who took to the country life like a duck to water. They were soon planning their gardens, most of them never having had an area to garden before in their lives.

We soon got enquiries about the possibility of hiring the Flymo and the rotavator, and a new business was born. It became evident early on that one of each would not be enough, especially if we had any breakdowns, so we went out and bought another

Flymo.

The matter of the rotavator was different, as they were much more expensive and we could not afford a new one. Where could you buy a second hand one? They were not exactly common to begin with.

Here, once again, Lady Luck stepped in. We spotted an advert in the paper for, believe it or not, another Merry Tiller. We quickly phoned the number and the man who answered insisted that he would bring it round to us. He duly arrived and told us that he had bought it to do his own new garden and having completed the work, no longer required it.

We quickly reached an agreement on price, which was much lower than we had expected and handed over the money. He then suggested that we should drink to the deal and said he had just the thing in his car. He arrived back with a full bottle of Canadian whisky. It seemed that he was a pilot flying the route to Canada and back and he bought a case of this each trip. It was excellent whisky, quite different to Scotch and smoother in taste. We settled down and what with one thing and another, it was soon midnight and we had drunk the whole bottle between us.

He insisted that we should accept another bottle from him, as we had mentioned what fine whisky it was and he then set off home. We hope he made it. We, in the meantime, had recovered about a quarter of what we had paid him for the rotavator in whisky, a wonderful bargain.

Chapter Twenty-Five

In June, we once again sallied forth for a holiday on the Thames. This time we decided to go westwards from Oxford, to the head of the navigable section at Lechlade.

This was a mistake, as we soon discovered. Nigel was four years of age and we had a baby on the way but the stopping places were far and few Judy, a pregnant lady, needed the breaks. On top of this, pubs and hotels were not inclined to allow us to stop for a meal with Nigel in tow, unlike the ones east of Oxford who had welcomed us with open arms.

On one day, we had already stopped at two pubs and been turned away. It was almost two o'clock when we tied up to go to a third pub. A tired and fed-up group filed into the bar. Yes, they did do meals, but the child could not stay in the bar.

At this, Judy blew up. 'Where can we take a child to get a meal? This is the third pub we have been to, and we want something to eat.' The publican relented, but asked us to go into a back room. We did, and had a meal, but we did not enjoy it.

As soon as we had got up to Lechlade, we about-turned and hightailed it back to Oxford and the east of the river, where at least we knew we could get a proper lunch. I only hope that things have changed since then, because it must have put off a lot of tourists from visiting that area.

The boat we had this time had a petrol engine, whereas the previous one had been diesel. I had forgotten that diesels are more economical than petrol. On the last trip we did not have to fill up even once and so I was expecting to do this trip on one tankful. How wrong can you be?

As we came downriver towards Sonning Lock, the engine gave a small splutter and we found ourselves floating majestically, without an engine, down the centre of the river.

'What happened?' said Judy.

'We've run out of fuel,' I said, after checking the gauge that I should have looked at before.

As we turned sideways on to the river, the riverbank seemed a long way off.

'You're the swimmer,' said Judy, 'you'll have to get in and pull us to the bank.'

'Where are my trunks?' I said.

'We've no time for that, you will have to go in nude,' said Judy.

I stripped off, hoping that no boat would come along to see me au naturel.

Just at this moment Cliff saw another boat approaching and proceeded to hail her.

'Wait while I get something on,' I said. Too late. The other boat swung alongside. They pretended not to notice my nudity. I meanwhile, tried to place my hands in a strategic position, before dashing into the cabin to try to make myself respectable.

Emerging from the cabin, I found that they had taken up a tow and we were heading downriver to berth at a boat yard where there was fuel. We thanked our saviours profusely and vowed not to run out of fuel again.

Luckily, we had no further mishaps and returned to Oxford after reaching Henley, our furthest point downriver. Notwithstanding the problems we had faced, we still enjoyed the holiday overall. It was a pity about the pubs on the river to the west, as it was beautiful countryside and well worth a visit. But it was back to reality.

At the beginning of July, Neta retired to the potting shed, obviously going into labour. Judy, once again, got herself totally involved, spending the better part of the night with Neta, and helping to deliver eight beautiful puppies.

Yes, they were golden labradors, just like the 'Andrex' adverts. Surprisingly, Neta turned out to be the perfect mother, feeding, tending and scolding them and keeping them well in check.

We had no difficulty in finding homes for them, although we were tempted to keep one of them, but two of our own were enough. We were kept informed of their progress by a number of the new owners. One trained his as a gun dog and said it was the best one he had ever had. Well, we always said we produced the best, now it seemed this included dogs.

On the other front and what a front, Judy started to get huge – so much so, that the doctor was sure it would be twins. Twins! I ask you, as if there wasn't enough livestock about the place without the hassle of twins.

But, as usual, Judy took it in her stride. Mind you, she did have to have some help with her shoes and washing her feet: one, she couldn't see them and two, she couldn't reach them.

Come September, when the baby or babies were due: nothing. It must be a boy or boys, they were usually late.

It took until the beginning of October for anything to happen and then the doctor was thinking on the lines of inducing the birth. But just as we were facing up to this prospect, at three weeks overdue, Judy's waters broke. At least this time it was at a more convenient time; six in the morning.

I was in my usual panic, but Judy was calm and we set off for the maternity home. I was once again shuffled off elsewhere to await the happy event.

When the phone finally rang, I was informed that we had a girl. What happened to the twins? Even Matron was expecting a second and had kept on urging Judy to push after the first had appeared, but there was nothing but water.

Shooting round to the home, I was met by the sight of Judy cuddling what looked just like a doll. It sported a head of black hair (this later changed to blonde curls). She was beautiful and we called her Catherine Ann. The Ann was a tradition in Judy's female line and we liked the old-fashioned Catherine with a 'C', not the modern spelling with a 'K'.

What I want to know is where did the blonde hair go? As she got older, she got darker. Not that I am saying that was a bad thing, just a puzzle, Nigel's hair stayed red to this day.

We decided that two were enough, especially as we had been blessed with one of each sex.

It reminds me of two other stories about family size. A friend of ours tried repeatedly for a number of years to have a boy, but eventually landed up with eight girls. They were urged to try again for as one wag said, 'You could have a netball team.'

The other was about a young couple who bought the cottage

next to us. They were always adamant that two children were enough for anyone. But after having a boy, they followed him with twins. They did get their legs pulled. It's a funny old world.

Chapter Twenty-Six

Now we had two Merry Tillers and two Flymos, the business of hiring them out started in earnest. The new gardens on the Gamesley Estate, together with new housing estates in the rest of Glossop and nearby Stalybridge, were the main markets.

I wound up in charge; delivering, giving instruction and collecting on completion. I also had to make repairs when necessary, which were rather more often than we imagined. It was amazing what people used them for and did to them.

The Flymos were used on areas which were little better than tips, causing dents in the blades, which needed resharpening or replacing. As the blades were connected directly to the crankshaft of the engine, this harsh treatment finally bent the crankshaft of one machine, making it unusable.

We eventually decided that we would discontinue hiring them out, as it was too expensive. Although we had numerous enquiries and people pleaded with us to hire again, it was too much of a chore and took up a lot of time, which we did not have.

The Merry Tillers were much more robust, but still had to be maintained. One had to have a new piston and rings when a customer tipped it completely over, allowing all the oil out. Despite this, he had continued to use it until it seized up. I did not find out until I came to start it again. The customer had never said a thing. Naturally, I went round to remonstrate with him, when he baldly stated that it was all right when he had finished, it must have been someone else. It was my word against his. We started asking for a deposit against damages from any future hirers.

A couple of tales spring to mind about the hiring of the rotavators. The first was the chap on the Gamesley Estate who, having dug up a cable, bent down to inspect it, leaving the machine in gear to continue on its Merry Tiller way. I arrived to collect the machine just in time to see it hit a paling fence on the boundary. It proceeded to smash the palings over a distance of some twenty-feet before he had the presence of mind to stop it.

On another occasion, I received a call saying that the machine would not start. Arriving at the address, some four miles away, I discovered that the chap had started to dismantle it to try and find out what was wrong. As he was doing this on the patch of earth he had just rotavated, he had lost most of the nuts, bolts and screws and I had to lend him the other machine. He definitely lost his deposit, despite his protestations. You do get them!

But all in all, we did manage to make a small profit and more to the point, anyone who had rotavated their garden would return to buy our other products. It was all good for trade.

Another change in our business came about this year. We had, over a period of time, established good contacts with a number of wholesalers, but these were limited. This year we were approached by a grower from Holland, who could match and even undercut our normal suppliers.

We decided to take the plunge and ordered a load of trees and shrubs, particularly Azaleas and Rhododendrons. They, as usual, arrived in the autumn, to be paid for in spring. The lorry that delivered them was so big that, after three attempts to reverse into the drive, the driver gave up. We were holding traffic up both ways and expected the police to arrive at any time. We had to go down the side road and manhandle the bundles through the back yard. It was a very tight fit, but we managed.

The stock was of a very good quality and we continued to use the Dutch suppliers from then on. We did have fun with the bills, which were in guilders and caused some headaches at the small local branch of Lloyds. I wonder how they will deal with the Euro?

Chapter Twenty-Seven

At this time of the year we had a new diversion. Advertising comes in many forms: leaflets, local and national newspapers, radio, television and even the cinema screen. We only used the local newspaper and then very sparingly. Obviously, there was word of mouth, but we used another one, purely by chance, but with mixed results.

One of the Local Authorities adjoining Glossop, before re-organisation, was so small that it did not have its own Parks Department, using the part-time services of the one in Glossop. Once a year, Local Authorities had a major annual function, the 'Mayor Making', which dignitaries from the surrounding towns attended. It was the civic highlight of the year. Most towns could put on a splendid show using their large town halls, bedecked with floral arrangements and opulent banquets. Longdendale, as it was called, being so small, could not run to these grand affairs and had to use a small hall and put on a buffet and drinks, none of the splendour of its neighbours.

The Clerk to the Council, who was only part-time (they could not afford a full-time official) was a customer of ours and had become a good friend. One day he suggested that we might put on a display for them in the hall, on stage. As they could not pay, they would allow us to display a sign saying who had carried out the exhibit. We agreed, feeling that it would be a good advert for us.

On the day, we bought in cut flowers and using our own stock, including potted houseplants, we set to work to create a magnificent exhibition. Although I say it myself, it was truly a sight to behold. After all, Cliff had done this very task when he had worked in Hyde Parks Department. The Clerk to the Council was well pleased with our work, and it was agreed that we should come back around eleven at night to remove the plants and clear up the room.

When we arrived, most of the people had gone. Imagine our

dismay when we saw that all the cut flowers and a large number of the pot plants had disappeared. The Clerk to the Council was most apologetic, offering to compensate us for the loss. We declined; they could not afford it.

It seems that the good burghers and citizens of the adjacent boroughs had taken the cut flowers, thinking they would go to waste. The pot plants had gone when things had got out of hand, and nobody had stopped them. It really comes to something to know that Mayors and Mayoresses, Councillors and their wives, were nothing but grabbing thieves. We consoled ourselves that at least the plants were good quality and maybe they had noticed our Garden Centre's name. Perhaps they would come and visit, but no way would they get it all for free then.

The next year we were asked to do it again, but this time we received official invitations, on the premise that if we were there in person we could defend our products and personally advertise our wares. Rather reluctantly, we agreed and turned up in our finery. It was quite a do, but to us a bit boring.

Now, I have to admit that the cut flowers were superfluous to our needs. It was the other plants that had been, let's be honest, stolen the previous year. That was what had annoyed us. So, at the end, we announced that the cut flowers would be given away, but would they refrain from taking anything else. It worked, as we did the handing out ourselves and we even got an official thank you for our time and trouble. We continued to decorate for this function every year until the reorganisation of local government in 1974, when Longdendale was merged with eight other authorities to form a new authority called Tameside (without the 'h').

Mind you, the thefts did have their compensation. Possibly the Council were trying to make up for what had happened. They broke protocol by offering us a planting contract that no one else had tendered for, I don't know what the district auditors said about that.

The contract was for planting of an area adjacent to the old courthouse, the front garden of an old peoples home and a plot of land surrounding a car park. It was a big contract for us, hard work, but very much welcomed.

I think we would have been better off with a pneumatic drill than a spade and pick, the ground was so hard and stony. Unbelievably in this barren ground, all the trees, shrubs and roses that we planted survived, we did not have to replace a single one.

This was not the first time we had carried out a planting contract, although that time we had not intended to do the planting ourselves. The firm Roneo had opened a new distribution centre just up the road from us and a condition of the approval for planning was that they would plant up the surround to the front and side and lay lawns. We were offered the job, as we were so near and since they didn't know anyone else in the area. We were not really into this type of work but we were loath to turn it down.

I knew that one of the Parks Department managers did some gardening work on the side, in his spare time. I asked him whether he would be interested in doing the work if we supplied the stock. After inspecting the site, he said yes and gave us a price for his part. We added on the stock cost plus a small profit and submitted a figure for the total work. It was accepted and work was due to commence in two weeks time.

On the day, our friend turned up with his crew, consisting of the Parks Superintendents of Oldham and Shaw, together with two other Parks men. They stuck in to attack the ground and at lunchtime they stopped, went off and never came back.

Ringing up my colleague in Parks, I asked what the hell were they playing at? He was most embarrassed and told me the ground was too hard and his co-workers were not prepared to continue. Of course there would be no charge for the work they had done already. What work?

He might well have been embarrassed. They had done precisely nothing. But we had given the price, not them, so we would have to do the work ourselves. We inspected the site. Most people who have bought new houses would appreciate what we found. Solid clay, bricks, stones and all kinds of sundry building materials, only our site was ten times the size of a new house garden.

We started in with picks and shovels, breaking up the ground and removing the debris. This was where the trailer we had purchased with the Merry Tiller came in really handy since we had mounds of bricks and stones to move to a tip in our grounds. As you remember, we had no brakes on this contraption, so we had to time our runs to miss any traffic and the police, as we did not have a licence to travel on the highway. Luckily, it was only a short distance from the depot to our driveway.

We had had to take a week off from our normal work to do the job and we were not sure whether that would be enough. We worked from dawn to dusk and just managed to finish by the following Sunday.

We cheated on the lawns though. Firstly, we had no supply of turf and secondly, we were not confident that we could lay turf, especially on that ground. We contented ourselves with raking the surface to a fine tilth, sowing grass seed, which incidentally we sold, rolling it, watering it and hoping for the best. Wonder of wonders, it produced a marvellous sward. At least we did not get any comebacks from the firm – and we got paid!

Chapter Twenty-Eight

Winter was now the traditional time to do house decoration, since we did not have time during the rest of the year. This winter was no different and we embarked on the refurbishment of the hall and landing.

Both areas had panels surrounded by a timber beading, the size, varying according to where they were. Some were eight foot wide and some only two foot wide. We painted the background light beige, the panels themselves in cream and the beading in a darker beige. Together with the skirting, it looked fine. Once again it took for ever, but the place was ours and we were determined to make it look agreeable; a home that looked lived in.

Next we had to tackle the stairs and ceiling, not something to be done lightly. This was going to be difficult. Remember, we were talking of a ceiling height of twelve feet six inches downstairs and ten feet six inches upstairs. This made the stairwell a height of over twenty-three feet. We rigged up a system of ladders and wobbly planks and stood by for vertigo!

Cliff ventured out onto the plank. He was soon back. It was a long way down. I tried it and managed to paint some of the wall, but it was quite daunting. I could only do it a bit at a time, with Judy hanging onto my legs in the vain hope that if I did stumble, she could stop me falling.

We managed somehow, even Cliff bravely attempting some stints. Next time, God willing, we would call in the professionals and let them worry.

Apart from the three front bedrooms, which we did not use, and the study, we had now decorated the rest of the house. This was really going to be a Forth Bridge job.

We had also installed a free-standing boiler in the old hearth of the kitchen, which supplied the hot water and kept the kitchen beautifully warm. We still did not have any central heating and continued to warm only those rooms we used downstairs and had a small electric heater in the nursery room upstairs for Catherine. Nigel would just have to make do with plenty of blankets.

There was also maintenance to do outside. The potting shed roof needed refelting, as it leaked. In order to ascertain how much we needed, I decided to measure it with Nigel's help.

With Nigel on the roof and me at the other end up a ladder, the length was easily measured. Now came the width. The roof sloped to the front, so I took the ladders round to the highest side and passed the tape to Nigel. He backed down the roof, whilst I gave instructions. Now, due to the height, I was unable to see the front edge, so I continued to pay out the tape, whilst Nigel backed down the roof. All of a sudden there came a cry and the tape was snatched from my grasp. Yes, you've got it, he had walked backwards, straight off the roof.

I rushed round and Judy dashed from the potting shed, expecting the worst, only to find him covered in dust and laughing his head off, none the worse for his fall. Naturally, I got the blame.

We also needed to paint the glazing bars to the greenhouse, but at the time we didn't know how to do it, so we chickened out. It would just have to wait for that mythical future we keep talking about.

Chapter Twenty-Nine

In 1970, the Gamesley overspill estate was really taking shape and we got a constant stream of customers. Unfortunately, we got one customer who was not so welcome, removing some five dozen rose bushes during the night. We did find out who did it, as the police noticed a house with rose bushes planted about every six inches (not what any normal person would do). But we could do nothing about it, as they had been careful to take all the labels off them and we could not prove they were ours. It did give the police some help though, as they were able to keep a watchful eye on them. As they inevitably carried out other burglaries, they were eventually caught.

We had other characters who made many visits, like the chap who, after collecting his dole, would go into the pub and get drunk. He would arrive in a sort of stagger and slurringly ask for a pot plant to take home as a peace offering to his wife. I don't know whether it did him any good but, as he came time and time again, it was certainly good for business.

Every Friday, when the binmen came, they parked their lorry on the side road, went into the pub for a drink and then straight across to the Garden Centre to buy plants. Judy, being the outgoing type, got on very well with them, cajoling them into buying some things they didn't want. But we always did them proud at Christmas, with drinks and a little cash bonus.

Judy always said, 'Keep the binmen, the postman and the milkman happy and you should have few problems,' and it seemed to work. I don't know about keeping the milkman happy though!

Then we had the two old dears, very genteel, who were into growing sweet peas and loved to talk to Judy about them. On one occasion, one said, in a whisper, 'When I have used my potty at night, I always water it down and it makes wonderful fertiliser.' I have heard it does, too, but who uses potties these days?

There was another character who was forever talking of his tomatoes and how wonderful they were. What we wanted to know was why he always came and bought tomatoes from us whenever we had them ready.

Now we had the estate we also got their kids. Being used to the city, they went slightly crazy in the countryside. This resulted in our first case of vandalism. Someone, unknown, tossed a couple of stones through our greenhouse roof. Little sods!

I talked earlier of the need to repaint the greenhouse and the problems we faced. Now we had no choice but to do something, so we worked out a method to move up and down the roof to both replace the damaged glass and repaint the bars.

This involved placing a plank on the roof attached to two ropes, which were slowly lowered down the roof, allowing work to progress in stages. Naturally, we required someone who did not weigh a great deal. Poor old Judy fitted the bill once again.

Ignoring her protestations we dispatched her up the roof to carry out the repairs and the painting. It took a long time and she needed frequent rests, together with the fortification of the odd drink. I think the drinks cost us more than the paint, but it was worth it.

Talking about damage reminds me that during the winter we started to get damage to trees by rabbits, who used to chew the bark off round the tree trunks, killing them. The trees I mean, not the rabbits! Urgent and drastic measures were needed. Through a friend, I purchased a single barrel shot gun, a Canadian Coley, which I was assured was a very good gun. Like the good citizen I hoped I was, I naturally obtained the appropriate licence.

It was indeed a good gun, as I was able to dispatch a number of rabbits and within a short time, the damage they were doing was curtailed. They made a tasty pie too, if you didn't mind having to spit out the shot.

There was another tale about the gun during this year. We had a number of fruit trees down the bottom end of the garden and these were subject to the usual scrumping. (For those that do not know the term, it means stealing apples.) While we only had the village lads, this was in reasonable proportion but now we had the estate kids as well, it was getting out of hand. It wasn't the loss of the apples that was the trouble, but the damage done to our stock whilst they were trying to get at the apple trees.

One day, Judy noticed a crowd of youths about to descend on the fruit trees. She went out onto the terrace and shouted at them to 'b... off!' This, unfortunately, had no effect. In fact she got a real mouthful of abuse from them. This made her mad and returning indoors. she picked up the gun, loaded up and returned to the terrace. The lads were already over the wall and about their business filling both their pockets and their mouths.

Once again she shouted a warning, but to no avail. Pushing the gun up to her shoulder, she aimed up into the air and let fly. It had a miraculous effect. They swarmed back over the wall as if there were a pack of wolves behind them.

We learnt later from the woman who had the newsagent's shop nearby, that they arrived up there well out of breath and complained to her about what 'that woman at the big house' had done to them. She told them they deserved it, and warned them that Judy would brook no nonsense from anyone. This had the desired effect and we saw no trace of them again.

When we learnt about her actions, we were a bit worried about the possibility of the police descending on us. They would not take kindly to someone discharging a gun at somebody, let alone children, no matter what the provocation. But, presumably, the newsagent's words had deterred them from saying anything to their parents and we heard not another word.

However, we were not entirely convinced that they would not come back, so we took further action. If any mother in the village was presented with a child covered in a sticky black concoction, which would not wash out, I wish to apologise. This was a foul mixture of car grease mixed with carbon black, guaranteed not to be messed with, which we liberally spread on the inside top edge of the wall. It certainly did the trick. We saw neither hide nor hair of any scrumpers again.

Chapter Thirty

With the advent of the new estate, we started to get enquiries about the possibility of using a room in the house from the local interest groups who were setting up.

We had had an enquiry the year before from the Women's Institute, but turned them down because we thought we had enough on our plate. Now we were approached by a group who wanted to set up a Senior Citizens' Group and this led us into thinking that perhaps we could let the lounge for an occasional meeting. Every little helps and the house was not paying for itself just acting as a home.

So we agreed, but pointed out that we did not have any chairs available.

'No problem,' they said. 'We can get hold of those.'

So it was arranged.

As the evening of the first meeting approached, the organisers came round to finalise things. The chairs were to be delivered around half past five and the meeting would commence at six thirty.

With the two lady organisers present at just twenty past five, we awaited the arrival of the chairs. By ten to six with no sign of the chairs, we were getting a bit agitated, as we had to set them up.

Just before six we had a phone call to say that the van bringing the chairs from the local library had got stuck under a low bridge and what should they do? Jumping in my van, I shot down to the town where the other van was stuck. I was met with the sight of two police cars and several spectators, all arguing about the best way to get it clear. The van was well and truly stuck.

I didn't have time to join in, as I needed to get the chairs back to the house as quickly as possible. The old people would be arriving shortly with nothing to sit on. The threat of a few dozen pensioners sitting cross-legged on the floor, as if round a camp-fire, demanded urgent action. I made five trips backwards and forwards, arriving with the last lot just as the folk arrived.

Meanwhile, they had let the tyres down on the other van, and were inching it out forwards from under the bridge. They had to do it this way as the road sloped downwards and the van had approached from the top end where the height was least.

It turned out later that the authorities had been constantly resurfacing the road over the years and quite forgotten that the measurement shown on the bridge was now wrong. It was amazing how quickly that sign was changed and another was erected on each side to give extra warning. This was no consolation to the driver of the now truncated van.

Meanwhile, back at the house, the old folk were arriving in their droves. The organisers and we had no idea how many were going to turn up. All we did was open the door and push them in. It turned out later that eighty-four people had turned up for the inaugural meeting. It must have been like the Black Hole of Calcutta in there, but no one complained. If all the meetings were going to be as well attended, there was no way we could accommodate them.

Luckily, at future meetings, there was a much reduced number but, unfortunately, within six months they had found alternative premises.

The success of this venture made us think of what we could do with the underused rooms in the house. We decided to offer the lounge to other bodies and to open it up as a cafe at weekends for our customers.

We approached a builder to convert the large window in the lounge to a door and to refurbish the former butler's pantry next to the kitchen. This work was done over the winter to be ready for the spring onslaught. Meanwhile, we scoured the shops for chairs and tables to set out in the room, together with all the paraphernalia required to serve cups of tea and snacks. This formed quite a dent in our finances but we felt it would be worthwhile. Roll on 1971.

Chapter Thirty-One

1971 was a very eventful year. Not that the previous years had been without their incidents, but this year seemed to have even more crowded into it.

Judy wanted to increase the livestock by farming pigs. Cliff and I definitely put our foot down on this idea. A friend of ours who was six foot two and solid with it reared pigs and even he had recently been gored by a boar. No way were we going to allow her to look after pigs on her own. Goodness knows what would have happened if she was attacked. Even a sow weighs a damn sight more than Judy, it would be no contest.

I think she took umbrage at this, because she decided that she would take on breeding rabbits and hamsters instead. We could hardly object to these. Not many people get gored by a rabbit! So we set up hutches and cages in the bottom garage, purchased two doe rabbits and a buck, also one each of the hamsters and installed them in their new homes.

It was not long before we realised that hamsters have very sharp teeth. They quickly chewed their way through the wooden cage. In desperation, we set them up in a large biscuit tin with a wire mesh front. Get out of that if you can! What we didn't know was that hamsters are rather susceptible to the cold. Now, Glossop was not noted for subtropical weather and sharp frosts in early spring were the order of the day.

Before long we got a rather exceptional cold spell. It even froze up the water in the hens' drinking troughs. One morning, Judy found the poor hamsters frozen solid. Metal tins were not obviously the best things to insulate against cold. This obviously put an end to the breeding of hamsters, this pair in particular.

As for the rabbits, after six solid months they were still not breeding like rabbits should do and as this was supposed to be a commercial endeavour, after a lot of argument, Judy was persuaded to abandon this venture also.

Typically, six weeks after selling them, one doe produced six young and later the other did the same. Just our luck!

Chapter Thirty-Two

The cafe was now ready for opening and we had recruited a friend to organise the food and drinks. I would collect the bread and muffins early on Saturday, Judy having already got the ingredients, baked scones and cakes during the week. We had also purchased a massive freezer, as we could not be sure that we would have the customers to consume everything we provided, especially when the weather was not up to scratch.

What all this did produce was the most delicious selection of sandwiches at lunchtime for Cliff and myself. It was very much appreciated, although not sound economics.

It did not take long for Judy's jam and cream scones to catch on. We had quite a number of people who came just for them, but we did usually get some trade in the garden as well.

The hired room was also much used by bodies of people such as the Women's Institute and the various special interest groups setting up on the estate. On one occasion, the local Conservative Party hired it for a meeting with the local MP. Cliff was late home this evening. It was raining and he swept into the drive, unaware of the impending meeting, just as the MP got out of his car to cross to the door. We nearly had an ex-MP for the High Peak constituency but, nimble as a fox, he leapt sideways into the porch just in time.

Mind you, this was not the only incident concerning Cliff and his driveway. One morning, as he was about to set off for work, and you have to take into account that Cliff only wakes up properly at about ten o'clock in the morning – in fact we believed that the car drove itself into Manchester on autopilot, he reversed out of the garage onto the driveway as usual and promptly collided with a car that had just driven into the Garden Centre. As Cliff said, 'One does not expect a car to be in your driveway at that time in the morning,' especially as the Garden Centre was not even officially open at that time.

Naturally enough, he had to admit liability, but it was not with any willingness on his part. After all that, the customer did not even buy anything. There's no justice in the world.

As the bedding plant season approached, we began the usual hectic runs to collect from the wholesalers. Bedding plants were one thing we were not cut out to deal with ourselves.

The Volkswagen van had been doing sterling service for a long time now and fitted out with staging, I could carry around fifty boxes on each run. Having done some three runs, I was back again at the nursery to collect another batch. A near neighbour to the nurseryman had just bought a Volkswagen caravanette and asked how I was doing with my van and whether it was reliable. I was able to reply that it ran very well, only needing servicing and the occasional new tyres over the five years we had it; no problems.

Starting back home, I reached a set of traffic lights at Wythenshaw, near the nursery. The lights were at red, so I pulled up and waited for them to change, attempting to put the van into first gear at the same time. There was the sound of churning gears and other sundry noises. Looking down, I thought I must have attempted to put it in reverse, so I tried again. This time the engine stalled.

By this time the lights had changed, but nothing, I did would allow me to get into gear. Getting out, I cursed the fact that I had tempted fate by declaring the thing perfect. I had to push it off to the side of the road with great effort. Two tons is rather daunting even on a level road. I then went in search of a telephone.

Remembering that we were in the RAC, I asked Cliff to ring them and explain the position and location of the breakdown. Before long, a van turned up and on investigating the RAC man discovered that the cable to the clutch had snapped. This ran from the front of the van to the engine at the rear and had a plate over it, which on a vehicle of this age, was rusted up solid.

The only thing for it was to tow it to the nearest garage for repair. I explained about the weight in the back and he was amazed when I opened up for him to see a veritable forest of bedding plants. Luckily, his vehicle was up to it and we duly arrived at the garage.

The repair was going to take two to three days and I had to explain to the mechanic about the weight of the plants in the back. I drove it onto the ramp, which groaned under the strain of several gardens' worth of plants. Amazingly, it was able to raise the vehicle without too much trouble and so the garage owner said he would be able to cope.

He might cope, but would the plants? I was forced to travel each day to the garage armed with a watering can to tend to them. A customer in the garage would have thought I was mad, as I proceeded to climb into the van on the raised ramp with numerous cans of water, liberally sprinkling the sad-looking and wilting plants.

When we finally got the van back and drove it home, we were relieved to note that due to my diligent watering antics we had not lost one plant. However, as a result of the shenanigans, they were the most expensive bedding plants we ever had.

Chapter Thirty-Three

Catherine, my daughter, was now two and a half years old and since she was two, had continually asked that she be allowed to ride a horse. Where she got this urge from we will never know but by spring of this year, her constant moaning had goaded us into promising that if the lady at the local riding stable agreed she was old enough to ride, we would allow her to take it up. We fully expected the lady to say she was too young and that would settle that.

How wrong we were. The lady took several moments surveying Catherine and her diminutive stature and said, 'no problem,' and promptly kitted her out with a hard hat and a small Shetland pony. We were stymied and had to agree to her going to the stables every week for a riding lesson. Catherine loved it but we were not exactly over the moon and this went on for quite a number of years. Why do little girls fall in love with horse riding?

Spring also brought the usual rush of customers, not that we were complaining. It resulted in our first traffic jam. There were so many cars trying to get into the driveway that we reached gridlock. I had to spend a good twenty minutes directing traffic to unblock the mêlée. We also had to produce a notice to advise customers to park on the side road if there were going to be too many cars in the drive. Still you got those people who thought it was their right to go wherever they wanted creating the inevitable fracas. I would then have to spend more time undoing the harm.

Another little gem occurred at this time. While I was carrying a consignment of plants up to a customers car, I was amazed to see a group of children on our front lawn, complete with a tablecloth laid out with a picnic and a dog tied up to one of our trees.

On remonstrating with them, pointing out that this was private property, I was told that they had thought it was a public park. They had meant no harm but I still had to send them on their way, as if we let this happen once, it would no doubt happen again. They were quite abusive as they left. I didn't know children knew language like that!

We prided ourselves on our service, making all sorts of deliveries around the area, from small plants to larger trees. It obviously cost us to do this, but the resulting customer satisfaction made it worth while. We were into customer service before the term was invented.

Luckily, most customers were able to take their purchases away with them, although sometimes I wondered whether the driver could see where he was going when trees were pushed through open roofs or foliage was packed around the driver's ears. Still, they managed somehow.

This Easter, we had a family who were visiting relatives in the area. They were very taken with the varieties of rock plants we had and chose three dozen of them. As they were still going on to visit other places, they enquired whether we would be able to send the plants to their home address. They would pay the whole bill including the delivery charge, COD.

'Where do you live?' we enquired.

'The Lake District, Ambleside,' they replied. They would be going back the day after next, but would not be back this way.

Talk about sending coals to Newcastle. But who were we to object? Not that we had any idea how we would go about this. We took their name and address and promised to send their purchase off as soon as possible. Of course, we charged accordingly.

The next day we discussed how we could make the delivery. Parcel post? By rail? We would have to make enquiries and quick.

Just at this moment, the front door bell rang. It was a friend of Cliff's, Frank. He asked what was going on, so we told him about our dilemma regarding the transport of the plants.

He laughed and said he was going up there on holiday the very next day, so he could make the delivery. We asked if he was sure and would he take something for going, but he was adamant that he was going there anyway, it wasn't out of his way and refused to take a cent towards the travel. That had solved the transport arrangements and we didn't have to arrange delivery this way ever again, so we never did find out the best way to do it. Nice and cheap, though!

The upshot of it was a delightful letter expressing their gratitude for the fast delivery and amazement at the method. I am sure that they must have thought this was our normal way of dealing with long-distance deliveries and we weren't going to disillusion them.

Their relatives in the area made numerous visits to us to buy plants and never failed to tell everyone who would listen how we had sent those plants to the Lake District. It was all good for business.

Chapter Thirty-Four

That year we took a trip once again to the Thames, but changed our start point to Bray, where the Vicar came from. This allowed us to make it to the lowest section of the Thames navigable by hire boats, Teddington Lock. Below there the river is tidal and hire craft are not permitted. We took our gardener's son Peter with us, as he was of a similar age to Nigel. The weather was not very kind to us this year, being wet, blustery and with lightning at times.

We stopped off to have a meal at The Swan at Sonning, where we met with Judy's maiden aunt, Lily, who lived near by. We invited Lily to go for a short boat ride, although we wondered how she would be able to get aboard. We needn't have worried, as she hitched up her skirts and clambered over the side. (This was a lady in her seventies.) She really enjoyed the ride.

She was quite a character. When I had first met her, we had been staying at her home for a few days. She had asked us to play cards, having first got in some sherry and I was amazed to see her pour herself a large glass of sherry, don a green shade and proceed to deal cards like a casino cardsharp. She nearly always won as well!

Having dropped her off, we turned our attention to Peter, who had been trailing behind on the journey back from the restaurant and appeared very restless. He had a headache and could not concentrate. Had he eaten something that had disagreed with him?

It was only when he described his feeling of disorientation and that on shore, things seemed to move around him, that we realised he was feeling seasick, or in his case riversick. After a course of seasickness pills, he felt much better. Now he travels the world, but I think it's mainly by plane. We completed the rest of the holiday without any further mishaps, although we did get more than a little wet on occasions.

Back home, summer arrived with a bang or more accurately a deluge. It rained for days on end, the ground was like a Florida swamp and the customers were in very short supply.

One weekend, on a Sunday morning, I woke up at the usual time to see it once again pouring down. 'What the hell,' I thought, 'it's not worth getting up.'

So I lay there and tried to go back to sleep. But this idea was quickly suppressed by the arrival of Nigel in our bedroom. He wanted to know why we weren't up. I said it was because of the rain and that there was no rush, there was little we could do in this weather.

Going to the window, he looked out. There was a prolonged silence. Suddenly, he turned round and said, 'There's a lake down the garden, daddy.'

'Don't be silly,' I replied. 'It's only the rainwater running down the paths.'

'No, daddy there is a lake,' he said, insistently.

Struggling up, I staggered to the window and looked out. There was indeed a lake, covering half of the bottom part of the garden. 'Oh hell! There is,' was all I could say, frozen to the spot by what I could see laid out before me.

By this time, Judy had realised something was wrong. She joined me at the window and gazed in horror at the spectacle confronting us.

'Quick, get your uncle,' I said.

Cliff arrived and I have never seen him come round so quickly from his normal early morning coma in my life.

'What the…? Where's that come from?' he said.

'Don't say Bast, Unca,' said Nigel.

Cliff was not one for swearing, his usual limit was 'Blast it!' said in an explosive manner and Nigel was frightened by this and not wanting to see his uncle lose his temper and swear like this, developed this saying. But it wasn't temper this time, just amazement.

We dressed quickly and donned our wellies and raincoats, although fishing waders and sou'westers would have been more appropriate. We ploughed outside to see what had caused this debacle. All we could ascertain was that water was backing up out of a manhole near the branch railway line at the bottom of the garden and as the trackway was above our land, the water was not going anywhere.

Now, there was a culvert pipe, which went under our grounds into this manhole and then passed under the railway line into the woodland. It came out part-way down the hill and joined a stream at the bottom of a valley. An indentation in the ground near the embankment showed where there had been a collapse, blocking the pipe. As the water could not get through, it was backing up into our grounds.

What to do now? Somehow we had to get the water away, that was the first priority. We went round to the wood and I gingerly entered the culvert from its bottom end.

Now, this part of the culvert was made of brick and egg-shaped, some three foot high and two foot wide. Inching up it, armed with a torch, I saw that it was indeed blocked at a place that appeared to be directly under the railway track.

There was no point in trying to free it from that side. The resulting deluge from the backed up water in the garden would have swept me away like a tree in a raging flood. So, back to our side where we unsuccessfully tried to push a pole down into the indentation to release the soil blockage. It was no good, we could do nothing by ourselves and being a Sunday, there was no hope of getting anyone else to deal with it. We retired indoors and morosely studied the ever-widening lake, wondering what on earth we could do next.

Monday morning arrived. Peering out into the gloom, I was amazed to see that the water appeared to have gone down slightly. It couldn't just be because it had stopped raining, it was getting away somehow.

We all got up and went out to inspect it. It was definitely less deep, but we were still going to have to deal with it and preferably before it rained again. I informed my office and took the day off. Then I rang up a couple of contractors I knew in the civil engineering trade and one agreed to come round and have a look.

An hour later saw him, cup of coffee in hand, inspecting the area around the indentation. As the water was now below the level of the manhole, we prised the top off and looked down.

Naturally, all we could see was water but using a weighted line, we were able to discover that the bottom was some fifteen feet down.

Well, hang the expense, we had to get it sorted, so I gave the contractor instructions to start work to repair the damage. He was in a position to start work the next day, so I rang in and took another day off. At this rate I would have no holiday allocation left, but needs must. In hindsight, maybe I should have called in sick, I definitely felt it. That night we wrote a letter to British Rail explaining our position and laying the blame at their door, since, from what we could see, the blockage was under their railway line.

We went to bed with heavy hearts, knowing that this was going to cost us dearly. We were just about to be in a position to pay off our former debt and even make a profit for the first time since we started. We just crossed our fingers and hoped that it would turn out to be British Rail's baby.

Chapter Thirty-Five

The next day dawned cloudy, but rainless and we awaited the arrival of the contractor. He turned up at ten o'clock, which was not a very good omen for the progress of the job. 'The digger is due shortly,' he said, 'then we'll get started.'

A low-loader arrived with a tracked excavator on the back. They proceeded to unload it with much difficulty; either the driver was not very competent or the machine was very awkward to manoeuvre. By the time they had unloaded and got it to the dig area, it was virtually lunchtime, so nothing was going to be done until the afternoon. I could see my holidays disappearing before my eyes.

Unbelievably, they actually started after lunch. The contractor said, 'Leave it to us,' (now where had I heard that before?) 'we'll dig down slowly so that we can find out what has happened.'

I didn't like the word 'slowly'. It meant that the meter was ticking away while we still had no idea yet what action would be needed to put things right.

The man with the machine started to dig away the earth around the indentation. By teatime there seemed to be a fairly large hole appearing round the indentation, but nothing else could be seen. The water was still slowly getting away and the men decided enough was enough and promptly went home.

I went back to work the next day, but could not wait until I got home to find out what was happening. So I rang up after lunch to see what was going on. Judy confirmed that the man had eventually turned up to start work at half past nine in the morning.

He worked for only twenty minutes before the left-hand track of the machine came off. He had scratched his head and kicked the digger but no further work could be done, so he had gone home again. As I said before, this did not bode well for the future.

At home that night, we decided that the only way we were

going to keep track of the cost of the job was if Judy kept a diary of events. So we got an exercise book and started a day-to-day log of the goings-on. Judy was so incensed about the seeming incompetence of the firm that she took to the task with enthusiasm. It was perhaps just as well, as the future events that unfolded were to show.

It took them two days to get the track back onto the digger. The contractor promised that we would not be charged for the lost time. Like hell we would! In the meantime, we received a letter from British Rail denying all responsibility for the problem. As the saying goes: 'They would, wouldn't they?'

We now had to call on our solicitor and put him in the picture and what a picture. He ummed and ahhed, but could not give us any indication of the likelihood of our being able to pin the blame on British Rail. He said he would write back and suggest a meeting with them on-site to try to reach some conclusion.

British Rail wrote back suggesting various dates and we agreed on one when our solicitor could be present. Bye-bye went another one of my leave days.

British Rail arrived in some number, seven strong. It must have cost them a fair sum just to turn up. There was their solicitor, two engineers, a surveyor and three manager types. They looked at the situation, probed the dig, went round to the other end of the brick-egg culvert and generally said nothing. While all this was going on, the contractors man could not proceed: More expense! They finally left, informing our solicitor that they would be in touch. We were no further forward.

Two days later, our solicitor rang to say that they were still not admitting anything. They would only accept their involvement if there was proof at the end of the dig, which they had been unable to see, since the contractor had not hit bottom, so to speak.

Well we could only wait and cross our collective fingers that the final proof would be available once the contractor got to the seat of the problem. At the rate he was going though, this seemed many days away. At least the pond had finally disappeared.

Finally, on the Tuesday of the second week, the contractor announced that he had uncovered the problem. Judy rang me at work and I arranged to get off early to meet him on site.

Arriving home, I went down to the hole where Judy, the contractor and his man were surveying the scene. It was a big hole, some twenty feet across and at the bottom there was an eighteen inch diameter earthenware pipe set into the bottom half of the brick-egg culvert. Originally, the top half had been bricked up but, over the years, the mortar had perished and with the pressure of the recent heavy rain, this had collapsed in and the earth had followed it to cause the blockage, which had created our lake.

The problem was that the end of the brick-egg culvert was directly under the boundary fence between our ground and the railway line. Was it theirs or ours? There was no way of telling.

We rang our solicitor to inform him what we had found and to ask his advice. He told us he would contact British Rail but in the meantime he would check with our deeds to see if they shed any light on the matter. We crossed our collective fingers again.

At this stage the contractor could not do anything more until we had both decided whose responsibility it was and what should be done to rectify the problem. So off he went. Would we be charged waiting time? That was usual in the trade. How long would it be before anything could proceed? So many questions but no answers.

We had another visit from British Rail, only three suits this time. Did they think they were winning and didn't need the additional muscle? At least we all agreed what needed to be done, in fact they insisted on it. A new manhole would need to be built at the junction of the pipes. This was imperative regardless of who was going to pay for it. So the contractor, who had been invited along, was instructed to carry on. Farewell to more holidays.

Chapter Thirty-Six

The contractor started back on site two days later. His first task was to divert the water coming through our eighteen-inch pipe into the brick-egg culvert.

The system he devised would have done justice to Heath Robinson. He was going to block off the pipe so that the water rose once again up the existing manhole, then channel the water through a series of troughs over the top of the working hole and down through a pipe into the top of the brick-egg. The troughs would be held up by a combination of props, rope and just plain fresh air. Whether it would actually work was a mystery.

Mind you, the 'buggeration' factor that we seemed to have with contractors continued to plague us. The contractor's man asked if he could borrow our aluminium ladder.

'Of course,' said Judy.

She wasn't to know what he was going to use it for.

Off he trotted with said ladder. Judy was not watching on this occasion. Perhaps she should have been. It appears that he pulled it out to almost its furthest point and placed it across the twenty-foot hole, which as we know was fifteen foot deep. He inched across it to put one of the troughs in place.

Now we all have observed in the past the predilection of aluminium ladders to bend and that's when they are vertical. This one was horizontal. It really bent. And then it went further. It broke, depositing our erstwhile workman down into the depths. Something else broke: his right leg.

It was some time later before Judy realised that she had not seen the man for a while. Going down to the hole, thinking that maybe he had tootled off once again, she discovered him covered in mud and moaning.

Exit, one workman in an ambulance after a rather prolonged and slippery rescue operation. It was another stop to the job and what's more, we never did get paid for the broken ladder.

The contractor himself now appeared on the site. Either he was short staffed or he was very much a one man band (at least he was now!) Somehow he managed to fix up the contraption to take the water, but we had not had much rain for days and there was only a trickle coming down, which barely built up to the top of the manhole.

Two days later, however, it was severely tested. It bucketed down for three days solid. Naturally, no work was done over those days but the thing actually worked, sluicing water away and down into the valley.

Having made sure water was not going to invade the working area of the hole, the contractor set to work to form a foundation for the new manhole. Of course he needed concrete for this, he also needed concrete rings to build the manhole. Enter one lorry to deposit a concrete mixer, some bags of cement, sand and the concrete rings.

The road down to the bottom of the garden was now a well established surface of limestone chippings, which had received a further dose of compaction from the digger. There was however, no road from that to the dig. The lorry reached the end of the chippings and slowly tipped sideways in the mud, like a drunken man sliding off a pub bench, neatly offloading its cargo. At least that would be one job that wouldn't need doing. It took a further recovery lorry to tow it out and that only just made it.

You would think by now that the contractor would have learnt that there was a jinx on any job that had to be done around our establishment. But no, he just carried on manfully.

In order to lay the concrete and the rings, two men were needed. As the other was still off with his leg in several pieces, the contractor turned up with a new man. He was about the scruffiest fellow I had seen in a long while. I am sure he must have been picked up as a tramp from the wayside. We told Judy to keep well away from him, as he looked very unsavoury.

He may have looked scruffy, but he turned out to be a good workman. As a result, the work actually progressed quite fast and they managed to fit the manhole top and backfill round it in the next week. Some tidying up took place, but when they had gone, one broken ladder sat forlornly next to the new manhole and the garden looked like the Somme on a bad day: a reminder of the disasters that had befallen the whole muddle of a job.

In the meantime, our solicitor, after a lot of research, had not found anything to implicate British Rail in the failure of the culvert. After talks, it seemed that British Rail's solicitors could not find anything to implicate us either. *Quelle surprise!* We were told that if it went to court, no one could win and that we would both be better advised to split the cost between us and leave it at that.

We did not want any further expense it was bad enough as it was, so we agreed. Luckily, BR agreed too, but it still cost us fifteen hundred pounds, plus our solicitor's expenses, a figure sufficiently large to virtually wipe out a small, English garden centre. We would have to persevere.

Chapter Thirty-Seven

I was just waking from a deep sleep at around half past seven in the morning, the first rays of light penetrating our bedroom window, when there was an almighty crash.

'What the hell was that?' I said, sitting bolt upright. Judy stirred beside me and said 'Uh!' Sleep does tend to make one articulate.

The sound seemed to have come from the back of the house. Crossing to the window, I peered out. It was not fully light, there was a mist and I had not got my glasses on. All I could see was a goods train, stopped on the main line. That was not unusual though.

I was sure I had heard something so going back for my glasses, I checked again. I looked round the garden and over to the greenhouses. I would not have been surprised if they had fallen down, as they were old and had not had much maintenance. But no, they looked perfectly all right.

'What's the matter?' said Judy.

'I don't know yet, there was a crashing sound.' I looked towards the railway again. This time I could see it plainly and I realised what was wrong. The goods train had not just stopped, it had come off the tracks. There were wagons piled on top of each other like a pile of fallen bricks.

'There's been a crash. The train's left the lines. Should we ring somebody?'

As it was a Sunday, we couldn't think whom to ring, except the police, but it wasn't a road accident, so would it concern them?

We decided that the police would at least know whom to contact, so we rang them. We explained about the crash and asked whether they would pass on the message and send someone round to check if anyone was hurt. We got up and woke Cliff and Nigel. Nigel, of course, wanted to go and see immediately, but we said it was going nowhere and he could see it after breakfast.

After we had eaten, we piled outside and walked down the side road towards the tracks to see what had happened. The goods train was a double-header, which for those who do not know what that is, has two engines at the front. These were frequent on this line, taking coal from Yorkshire to the power station at Fiddlers Ferry. This was a twice a day run through the Woodhead Tunnel (which after a few more years, would be shut to traffic).

One engine was completely off the track. The second was still on it but the sudden stop had had a disastrous effect on the loaded wagons. Some had left the track, some lay on their sides and still more had leapfrogged the ones in front. Large sections of the track had been torn up as well. It was quite spectacular, but there was not a soul in sight and in the half-light of a cold October morning, the silence and the scene were quite eerie.

Nigel, like all little boys, was excited but we had to explain that although there appeared to be no casualties, there quite easily could have been, and that would not have been exciting, rather quite the reverse.

It did start the day on a high note, but since drizzle set in soon after and very few customers turned up, it was the only high note of the day. Later, trains and a crane turned up and after a great deal of pulling, shoving and lifting, the line was finally cleared. Next came a crew to clear up the coal and to reset the line. The whole undertaking took four days. This must have really upset British Rail's schedule as they only had one line taking not only the main line traffic through to Sheffield, but also the local traffic to Glossop.

Chapter Thirty-Eight

October was the month when we obtained our stock for the next year from the wholesalers. We had just received the stock from Holland and we were due to go down to Stoke for further supplies but, as before, we needed a lorry for this.

I had been able to borrow a lorry from a friendly builder for the last two years and once again I enquired whether I could use his lorry. They were quite amenable but explained that they had just got a brand new lorry and they needed to get some equipment to a site on the day that I wanted to borrow it. We agreed that they could have the Volkswagen whilst I took the lorry.

The day before, I took my ailing, old van round to the builders yard and picked up their spanking new lorry. I drove very carefully home, being especially circumspect of a single-track bridge, which was notorious for trapping unwary drivers due to its awkward approach and the right angled bend immediately after it. I survived the journey and hoped I would do as well the following day going down to the wholesaler's nursery, a trip, which would include some pretty narrow roads and bends.

The next morning dawned bright and clear, it looked like a good day to sally forth and bring in the goods. I set off early as it was a fair way and getting there, loading up, coming back and unloading would take up most of the day.

Driving through the next village, I became vaguely aware of a tightness in the steering but put this down to the vehicle being new.

After this, I had to negotiate a number of very sharp bends. The tightness got more pronounced and now included a slight wobble. Time to stop and check, I thought.

Pulling into a lay-by, I got out to see what was wrong. It was fairly evident: the offside front wheel was just about to fall off, as the wheel nuts were all loose. It was a good job I had stopped, a few miles further and the wheel and the lorry would have parted company.

I searched high and low for the jack and wheel brace: nothing! But I had to do something. I thought of ringing home to get the local garage out but that was sooner said than done. The nearest telephone box was two miles back down the road. There was nothing for it but to trudge back and hope that Judy or Cliff weren't outside and would answer the phone.

Luck was with me. Cliff answered, agreeing to get the garage out to me pronto. Unfortunately, garages were not always on call to go out to stranded voyagers, even if it was an emergency.

I trailed back, wondering if I would ever be able to collect our stock. If I couldn't, would the builder let me have the lorry for another day? Mind you, I had a good excuse, as it appeared to be his fault that the wheel was loose.

I had hardly got back to the lorry when up drove Cliff and the garage owner himself, armed with a jack of enormous proportions and a set of lethal looking tools. It only seemed a matter of minutes before the wheel was firmly fixed, although the garage owner was quite perplexed as to how the wheel had come loose on such a new vehicle. Bad maintenance was all he could suggest, even though the lorry was only three days old.

Setting off once again, I quite naturally, took it very steady. Every little squeak or tremble made my heart leap into my mouth. But the journey there and back was completed without any more trouble.

At home, we unloaded and I prepared to take the lorry back to its owner, with whom I would have to have sharp words. Maybe I should moderate them, as I might need to borrow the vehicle again. I drove off, hoping that nothing else would happen.

Arriving at the yard, I got out and went to find the builder. On approaching him, we both started to talk at the same time. He suggested I relate my bit, then he would tell me his news. I didn't like the sound of that. Anyway, I explained what had happened mentioning my displeasure.

He was quite shocked. 'It must have come like that from the garage. We haven't touched it yet,' he said. He would have words. What was his news? Where was my van? It appeared that the van was low on oil, I hadn't checked it, and neither had they. It had totally seized up over in Yorkshire. 'I have arranged to get it towed back,' he said by way of apology.

We both had had a difficult day, and no blame could be attached to either of us, he not having checked the wheel and I not having checked the oil. These things happen, but they always seemed to happen to me...

The van was not badly damaged. The local garage had got its second job in two days from us and offered to repair it at a reduced cost. We were back on the road.

Chapter Thirty-Nine

By now, we were due to replace one group of hens. However, the price of eggs had remained virtually static for the last two years, whilst the cost of food, replacement hens etc., was rocketing upwards. Even large producers were feeling the pinch. We, being very small producers, were on a hiding to nothing and there was no money in the kitty for replacements.

We decided that we could not afford to divert money from our other enterprises to fund the egg production and agreed to run this part of the business down. It had not been without its perks. It had paid for the van and given us an unlimited supply of 'free' eggs, sometimes more than we could cope with.

Around Christmas time a tragedy occurred. Whilst taking my usual long, relaxing bath one Sunday evening, I was suddenly brought back to earth by a loud hammering on the door. Judy shouted that I was wanted downstairs immediately. Poor old Sheba, our faithful labrador, had died.

Quickly drying and dressing, I rushed downstairs, to find Cliff and Judy pulling Sheba out from behind the chair where she normally lay and Neta, our other labrador, looked quite worried.

For some time now, Sheba had been off her food, but we had put this down to her and Neta as needing worming again. It seemed that she had just uttered a squeak, sighed and simply passed away. We took her outside, dug a large hole in the shrubbery and reverently buried her. She had been a very good companion.

Not long after, Neta started to show similar signs, so we took her to the vet. It seemed that she had an infected liver and that had probably also been the trouble with Sheba. There was no cure, so we took the agonising decision to have her put down. We did not want her to go through the same slow decline as Sheba.

That left us without the company of a four-legged friend. Judy was not so keen to get another dog. She felt that they were a liability when it came to going out or away. We, that is Cliff and I were more inclined to look at the companionship and protection that we believed Judy needed. Indecision reigned, but events soon overtook us.

As I have said before, Cliff worked as an Environmental Health Officer in the Moss Side area of Manchester and for some time now we had been friendly with a couple who had a pet and garden store. Cliff used to call in on a regular basis for a chat and to look for business.

On one of these occasions, there was a customer in the shop who had a golden labrador with her. She was enquiring whether our friends knew of any one who wanted a labrador. It seemed she had taken it in when its previous owner had wanted to put it down. He was not at home during the day and the dog, having nothing to do had been slowly tearing the cushions, carpet and chairs to bits. She already had a dog and the two animals were not getting on. Cliff was immediately taken with the dog and agreed to take it on the spur of the moment.

Arriving home, he left the dog in the car and came into the house to prepare the ground as it were. He announced to Judy and me that he had a surprise, neither Judy nor I had an inkling what he could mean, so he said he would go and get it.

Imagine our surprise when he arrived with a new dog. Judy was a bit put out. She was disappointed that we had not been able to discuss it. After the situation was explained to us, Judy soon thawed and Liz, as she was called, became a member of our household. She was to prove to be the best dog we ever had. Our initial worries about her tearing things up were soon gone, as she settled in. With people around her and plenty of ground to roam, she was not bored any longer, which was what had been the trouble all along.

We had considered using her as watchdog, but when we saw how friendly she was to all and sundry, we thought that was probably one attribute that she had did not have. We were proved wrong by two happenings the following year. The first was when the newspaper boy tried to go to the back door instead of the front. He came through the side gate whilst Liz was outside and was immediately pinned against the wall. Judy had to rescue him.

It seemed that if people came in the way they should, from the front and we greeted them, so did Liz, but uninvited guests at the back counted as an attack on the house. This was exactly what we wanted and she had not even been taught.

The second incident happened on a glorious summer day. Judy was in the kitchen where the food was prepared for the cafe, when one of our regular customers started to talk to Judy through the window. The dog was in the kitchen with Judy when the man outside, in a mock gesture, shook his fist at Judy.

The dog was up and running immediately. The man started running too when he realised that Liz was after him. He shot along the terrace and in one leap, cleared a gate I had recently put up to stop young children getting onto the main driveway. The dog was right on his heels.

Having cleared the man out of the garden, Liz returned happily to the kitchen. This was proof that if anyone ever threatened Judy, Liz would have them. Fortunately, the customer recognised the characteristics of a good watchdog and kept his hands in his pockets from then on.

As I became more involved in the nursery garden trade, I learnt more and more, particularly the Latin names of species and could recognise a number of regional variations on common names for various plants. Although I had always been interested in gardens and garden layout, I had never previously been concerned with plant identification. Now I was thrust into it and I had to know. This led to a number of problems, the worst being that I have a peculiar brain that decides to switch off when asked a direct question.

I could quite happily go round the garden telling people what this or that was, but if someone said, 'What's that plant?' Zap! Could I think what it was? No way. Yet, if I had been able to continue talking without being asked, I am convinced I could have told them what it was. I suppose there is a name for this condition but it was and still is, most embarrassing.

I could also embarrass other people, not deliberately of course. I remember Judy pulling me up when I had finished serving some customers. It appeared that, in describing a particular plant, I had inadvertently used an impolite word.

The plant in question was one of the conifer species, the *Pinus*. Apparently, I had pronounced it 'Penis', which is quite something else. It had Judy and Cliff and I suspect other customers in hysterics. Another one that always got them going was my way of saying *Lonicera Nitida*, I pronounced it 'lone y sarah nittyda'. Once I had been aware of this, I used to take great delight in emphasising the way I said it.

Mind you, there are quite a few oddments of pronunciation. How do you pronounce *Cotoneaster*? Do you say 'cotton easter' or 'cottone aster'? I prefer the second, but others the first.

I was also very fond of Spoonerisms: changing the starting letters of words round as a joke, like, 'I'm going to low the mawns.' Unfortunately, this sometimes happened by accident. On one occasion I was trying to be poetic and instead of 'shaft of wit', the inevitable came out, much to my embarrassment. Mind you, when it came to our hen manure, it was probably true, as Cliff could testify.

Chapter Forty

Another spring and as usual, things were quite hectic. Come Easter we were back to bedding plants. Now, bedding plants are big business to a nursery, as they bring in customers who usually spend on other purchases as well.

We had always prided ourselves on relying on word of mouth for our main advertising but this did need a push on occasions. So, for a number of years now, we had supplied bedding plants, at cost, to a children's home in the next town southwards, Marple, for their Spring Fair. Now, you might think that this was not good business, after all, we had transported the plants all the way from South Manchester, had had to water and look after them, and then had to transport them to the home.

Well, it pleased us to be able to help the children at the home, but also, with a sign saying all plants were supplied by us, it was amazing how many people came to visit us to shop at normal prices. They of course in turn spread the word further.

Usually, I had used the van to deliver on these runs, but it would only be half full. I had recently obtained a new Renault 16, which was the first of the hatchback combi-type cars on the market. In addition, the whole of the back seats came out, leaving a large open space like a van. I had set to and made racking to take twenty boxes of bedding plants. I drove it down to the bottom of the garden to load it up on the morning I was to take them to the home. Turning round, I drove back up onto the drive.

Judy had asked me to take our daughter Catherine, with me, as it would be a trip out for her and would get her out of Judy's hair for a while. Catherine was on the top terrace already, leaning over and opening the passenger door, I shouted for her to come. All she did as usual was to turn tail and run in the opposite direction.

Throwing the door open, I jumped out and began to chase after her. Suddenly, I heard a peculiar noise behind me. Turning, I was just in time to see the car moving ever so slowly, backwards.

I had left the brake off! But, as the car was on what I thought was flat ground, in my hurry, I had not expected it to move. I was wrong.

I dashed back and grabbed the bumper, as I was not able to get to the door. Digging in my heels, I tried to stop it going backwards. Some hope! The combined weight of the car, plus those wet boxes of bedding plants were too much.

I sort of skied down the slope, my shoes having little effect on those loose chippings. Partway down, the driver's door, which had been open, slammed into the wall and gracefully folded over on itself. The downward journey only stopped when the back wheels of the car hit a timber surround on the bottom flowerbed.

My shoes were full of small stones. I took them off and proceeded, in sheer frustration, to fling them as far as I could. It didn't really help, but it was something to vent my anger on.

Judy and Cliff came running over, having heard my shouting and cursing, stared at the car and me. Naturally, in my state, I blamed Catherine and Judy for the accident, not myself for being so silly as to leave the handbrake off. It had happened so quickly. Judy soon put me right on that one.

We inspected the car but, apart from the door being bent double, there appeared to be no serious damage. Even the engine was still running. With a great deal of pushing and shoving, we managed to bend the door back to a semblance of straightness, but of course, there was no way it would shut.

We had still got to get the plants to the home so after tying up the drivers door with rope, I got in the car from the passenger's side and drove off, without Catherine. I could not be held responsible for my possible actions and Judy was being particularly frosty towards me. I suppose I deserved it, you do tend to loose your rationality in such circumstances and say things you regret afterwards.

I got a lot of funny looks and questions when I got to the home but managed to parry them without making myself look a complete prat. It was rather chilly driving the car back and forth though, with the door hanging half off.

I happened to know a chap who did car repair work and he made quick work of making it good. I forgot to tell him I was intending to pay for it myself, as I wanted to avoid the insurance company taking away my no claims bonus. When I went to collect it he presented me with a full bill, believing that it was to go through my insurance. When I told him the situation he said that he could not reduce it, as the bill had already been put on his books. It was still less than the cheaper bill plus the no claims, but it was a blow all the same. Why is it that little children always seem to do the opposite to what you ask them?

Chapter Forty-One

This year we decided to do something different for our holiday. We still had to consider the dog, so while looking to see what was available, we came across a farm holiday just outside Leominster, pronounced 'Lemster', Herefordshire. This seemed just right for us, as we were not only taking a dog, but my mother-in-law too, and the peace and quiet of the countryside would suit us all.

The directions on how to get there were a little misleading and although the Abbey Farm, as it was called, was on the map, the nearest road leading to it, wasn't: it was a private road. We finally traced a way in from another direction, down another long, private drive.

The farmhouse was very old and three stories high. With its grandfather clock, brassware and log burning fire, it gave us all the atmosphere we could want.

Meals were gargantuan, this was certainly not a slimmer's holiday. After the evening meal, we usually settled down to play cards with our hosts, who introduced us to a game called 'Nap'. The Jones, whose farm it was, were a brother and sister who lived with their mother, who was well into her nineties. Even the brother and sister, who were unmarried, were in their late sixties. This did not stop them from being well ahead by the end of the week. We played for pennies and the game needed quite fast reflexes, as it resembled the game of 'Snap'.

One holiday episode does come to mind. Hereford was a market town and once a week there was a livestock market. Now we thought this would be good to show the kids.

We wandered through, observing the sheep, cows, even hens being auctioned. Then we came across the pigs. Judy took a great interest in these but remembering her earlier idea of keeping pigs, we had to hustle her along. We were not very sure she would not be tempted to bid for some and no way were we going to be lumbered with getting pigs home, let alone trying to breed them.

There was also the Leominster Agricultural Show, which luckily, was on during the time we were there on a beautiful sunny day. Nigel was very interested, as all lads are in the huge tractors and combine harvesters. Catherine was naturally more interested in the horses. We were all weary after trudging about all day, but it was worth it.

We visited many interesting places, like Hereford and Ludlow. The week passed quickly and on the day we were going home, we went to Leominster to get in the shopping for home. Who should we meet but Alan, my best man? We hadn't seen him for all of ten years. We had lost touch when he moved to a new job and we had no idea that he lived near Leominster, practising as an architect from offices in the town. We really did not have time to talk, but swapped addresses and phone numbers, promising to keep in touch. It's a small world.

Speaking of which, a colleague of mine at work went to Malta, where they met another couple. Talking about this and that, they mentioned that they came from Ashton-under-Lyne. Their new found friend said, 'Oh yes, do you know a chap called Peter Binns.' It seems he was a former colleague from Macclesfield. Unbelievable!

Incidentally, Alan has gone missing again, he is not at his old address near Leominster, perhaps if he ever gets to read this he might consider getting in touch.

Chapter Forty-Two

Mushrooms! No, this is not a gardening swear word. It was the next project we decided to attack. Those cellars were crying out to be used and would suit the growing of mushrooms we thought, having an even temperature all year round and little light.

First, we read up all we could about mushroom growing, then, having found a supplier of the material and equipment needed, we decided to go ahead. I set up a double row of shelves round the largest cellar under the kitchen, together with a central, island shelf. All this was done with second-hand timber from a demolition site. No point in spending a fortune. The shelves were three foot wide, with a front rail and the whole fitted with a plastic sheet liner.

We bought packets of growing medium. This was in the form of small, nut-like nodules, which would produce the thread-like fibres of the fungus, which in its turn produces the mushrooms above ground. We also bought the special compost, which when mixed with straw, created the bed in which the mycelium, which is what the thread-like rootlets are called, would grow. This 'glurk' when mixed, had to reach a certain temperature, before it could be placed on the beds. For this, we had to purchase a huge thermometer through the same suppliers. They had it all sewn up. The only things we had to supply ourselves were the beds, the straw and some topsoil.

Why is it that when you start something, everyone expects you to carry on with the rest of the job? Muggins landed up mixing the compost and straw himself with a large garden fork in one of the small cellars, although I did have help carrying the bales of straw downstairs.

Now we had to wait while it all started to react and create heat. During this time the whole lot had to be turned over every two to three days.

Two days later, I heroically heaved the whole lot over, having first checked the heat with that huge thermometer. It seemed to be doing nicely. Unfortunately, I wasn't. I woke up the next day covered in large blisters. I was allergic to the stuff. Hallelujah! I could legitimately get out of mixing again. Judy and Cliff took over for the next turning session, with the heat building up nicely.

About this time, a terrible pong invaded the house Was it the drains? We traced it to the simmering compost, which was putting out a distinct fog. We vowed that next time we would mix it somewhere else, for the time being we would have to put up with it. We got some peculiar looks from customers passing by and from visitors to the house. It was horrible. No mention was made of this smell in the instructions, naturally.

At the next mixing time: panic! It was not just heating up, it was close to combusting. The thermometer was almost too hot to remove. We had to do something. All we could think to do was to spread the mulch straight onto the beds to cool it down a bit and hope for the best.

We all pitched in even me, despite the likelihood of getting those blisters again, but this was an emergency. We added the topsoil as we went and then checked the heat. Yes, it was going down, we would check again the next day before we planted the mycelium.

Horror! On checking the next day, the beds were almost totally cold. We consoled ourselves that the instructions had said to spread on the beds when the heat was right. We presumed that the very act of spreading would cool it down. Holding our collective breath we planted the nodules and departed to await events. Soon we would have fresh mushrooms for sale and another profitable line. (By the way, my blisters did not appear, meaning I had lost a good excuse.)

We could hardly contain ourselves in anticipation of the first crop of mushrooms and kept sneaking down to check if anything was happening. But after three weeks: nothing. Letting it cool down must have upset things.

After four weeks, Judy came rushing into the sitting room, having sneaked a peek in the cellar. She announced that the mushrooms had finally arrived. Cliff and I followed her down to the cellar, fully expecting to see mushrooms sprouting everywhere. There, in one small corner, were about ten mushrooms. They had started growing, so the others would be along soon and we made plans to set up a stall that weekend.

In the meantime, we picked what had arrived, intending to have them for breakfast the next day; mushrooms don't last long as buttons if left. Anyway, more would be along soon, wouldn't they? They were delicious, far better than those that shops supplied.

By Friday, there was nothing. Where were they? The sign and basket sat forlornly on the kitchen sideboard.

As it was, we had fresh mushrooms for the next twelve months, at the rate of three quarters of a pound per week. Fine for us, but hardly a commercial success. They just kept popping up in various places, but only in small numbers.

Never mind, we did enjoy them, I was not going to get blisters any more and Judy and Cliff wouldn't have to mix that muck again. And we certainly weren't going to miss that smell.

Anyone who wants to buy a very large thermometer?

Chapter Forty-Three

Remember that shotgun? It had been very useful over the years in controlling rabbit damage and secretly frightening unruly kids. We had never, in all that time, ever taken the gun out of the grounds of the nursery. Guns are dangerous things and have to be licensed. Being, I hope, a good citizen, I had taken out a licence for mine right from the beginning.

There was one period when the Government decided to offer an amnesty to people holding illegal guns. As long as they handed them in, they would not be prosecuted. During this period, licences were suspended. After a year had passed, anyone holding a gun was to reapply for a licence. Like TV licences, a reminder went out to all who were known to have held a gun and I was one of them. Now, as you know (and I am afraid I have repeated ad nauseam), working full-time and running the Garden Centre did not exactly leave much free time. Quite simply, I forgot.

Near the end of the year, I received notice that I was to be prosecuted for not having a licence. I immediately tried to apply for one but was told that the court case came first and that the issuing of a licence would depend on the outcome. Attached to the summons was a form which said that if I did not want to appear in court, I could make my plea by post, 'guilty or not guilty', together with any extenuating circumstances.

Not wanting to waste too much of my time and obviously guilty, I decided to use this method. I sent it off and thought nothing more about it.

Naturally, the evening before and the day of the hearing, I was on hot bricks, wondering what the result would be and when I would hear the verdict. That evening, we retired to bed about eleven. We heard a verdict rather sooner than we thought. At a quarter past midnight, the front doorbell rang. Who was ringing it at this time of night? Looking out of the side bedroom window, I saw a police car in the drive. Had we been broken into?

Half asleep, I went downstairs, dressed in my pyjamas and dressing gown. I opened the door, to see a policeman standing there.

'What's the matter?' I asked.

'Are you Peter Binns?' he said.

'Yes, why?' I said.

'I have a warrant here for your arrest, as you did not turn up in court to answer the charges brought against you,' he said.

By this time, both Judy and Cliff appeared to see what all the fuss was about. Having heard the last statement, they were looking aghast at the policeman.

If anything, I was looking even more astonished. 'What are you talking about?' I said. 'I sent a form pleading guilty. The note on the form said I did not have to appear in Court.'

It was his turn to look put out. 'I'll have to look into this,' he said. Going back to the police car, he radioed in.

After a brief discussion, he came back. 'Yes, it seems that you did. You should never have been sent that form. All gun cases have to have the accused present in court. We will have to set up a new date for a hearing and you will have to attend.' So saying, he bid us goodnight and drove off.

I sat down on the stairs and breathed a sigh of relief. I had fully expected to be dragged off to prison and what would my boss have to say about that? Judy and Cliff were furious that the police and the court had made such a mistake. I could only reflect on what might have been.

We finally got to bed again at quarter to one. I had hardly got my head down, when the front doorbell went again.

'What now?'

I stormed downstairs, and opened the door.

There stood the same policeman. Had they changed their minds?

'I'm sorry, Sir, but I have to take the gun in.' I went and got it and received a receipt.

'Goodnight, Sir. I won't disturb you again.'

I should jolly well hope not, I told him in no uncertain terms.

Judy was hanging over the banisters when I returned, trying to hear what was happening. When I explained, she said, 'Why didn't he take it the first time?'

'I've no idea,' I replied and at this stage I didn't care. All I wanted to do was to get some sleep, but this did not come easily with the events of the evening rushing through my mind.

The due date appeared and I went down to the Magistrate's Court, feeling like a criminal. I had no idea what the procedures were and presented myself to an official, who was taking the names of the people there.

'Are you being represented?' she asked.

'No, did I have to be?'

'So you will be representing yourself?'

This was said in a rather disparaging tone, as though I looked like an idiot and could not string two words together. Perhaps she was right. I was told to sit down and wait till my name was called.

Looking around, I saw various people in discussion with what were obviously solicitors. Doubts now set in. Should I have had a solicitor? Was it such a heinous crime as to warrant one? Was I capable of presenting the facts so that they understood that it was only a lapse of memory? I forgot; was this adequate? It was too late now.

Courts are like hospitals, they seem to have a system whereby everyone for the whole day's business is asked to turn up at the same time. I suppose it is to make sure that they have a full programme and they do not waste valuable time. Solicitors must make a fortune while they wait with you.

Finally, I was told that my case was next but one and to be ready when my name was called. It was not long before this happened and I was sent into the courtroom, knees quaking visibly.

There was a case still going on and I had to wait until it was finished. I could not help hearing what it was about. The chap before the Bench was there for beating up a girl. This was not the first time that it had happened and he was well known to the police.

The Chief Magistrate conferred with his colleagues and gave the scoundrel a good telling off. 'Now, I thought, he will get at least twelve months in prison.' But no, he was given twelve months suspension on a surety of twenty pounds. In other words, as long as he did not do it again, he was not fined anything.

I thought, well, if they could deal with him that way, when the evidence was before them in the shape of a young woman, with a black eye and severe bruising of her face and arms, my case should be a doddle. After all I had harmed no one, apart from denying the exchequer my licence fee.

Now it was my turn. I was directed to the dock and gripping the rail, I tried to appear unruffled and confident, whilst at the same time trying to look the Magistrate in the eye.

Shuffling his papers, he did not even look my way.

'Are you Peter Binns of Gamesley House, Glossop?'

'Yes,' I replied.

'Would you like to say a few words?'

'Yes.'

(Of course, I wanted to add, but felt that would be too flippant.) So I explained what had happened and apologised to the court.

Mumble, mumble went the Chief Magistrate to his fellow magistrates on the bench. He had still not yet looked at me but after this exchange, he suddenly fixed me with a baleful stare and said, 'This a serious crime and I could fine you up to one thousand pounds and send you to prison for six months.'

I did not have one thousand pence, let alone one thousand pounds. What would Judy do if I was sent to prison? I was now visibly shaking.

He carried on. 'But we tend to believe your story and fine you twenty-five pounds. Step down.'

I think I stumbled down, rather than stepped. I was guided to another room, where I had to sign a cheque for twenty-five pounds. I only hoped that there was enough in my account.

But what of the gun? They had confiscated it. Would I get it back and would I get a licence? The police officer in charge settled that by giving me an application. When I had received the licence I was to collect my gun from the station.

As it was, once I had the licence, we decided that, as a new fence seemed to be keeping the rabbits out and we did not want to go through all that hassle again, we would sell the gun and be rid of the problem. We found a licensed dealer in Glossop and sold it to him. It was better to do it that way than to sell it to an ordinary person and possibly be in more trouble.

As for the charade of the court case, when one person, who had obviously caused damage and harm to another human being, could get off virtually scot-free and another could be fined for what was really a trivial offence, all by the same magistrates, it seemed to be question of your background. As an upright citizen who errs slightly, you are hammered by the courts, but if you are 'underprivileged' you can get away with attempted murder, not once, but several times. The current attitude of magistrates and judges, in the main, seems to be based on this principle. I am sure that the general public does not see it this way and it behoves them and the Government to sort themselves out. I know that if I were a magistrate, I would not act in this way, even if the Clerk to the Court were to suggest it. Yes, this is not politically correct, but I think many people would agree.

Chapter Forty-Four

We had now settled down to our yearly routine and entered 1974 with high hopes. The problems of the past years were behind us and we only had the loan to pay off. That seemed to be possible before the term, now that we were making a good profit.

Catherine was due to start school this year. Her horse riding lessons were still ongoing, although she had now moved up to a bigger beast. Naturally, she was now looking to get her own pony, but I was resisting this as best as I could. I knew who was going to land up looking after it; Judy and I. We came up with some very inventive excuses; the stable roof leaked and that would be an added expense. Unfortunately, she did not seem to be put off. We would just have to see.

The seasons rushed through without any great debacles or worries, life seemed sweet, with nothing to mar it. All too soon, Catherine started school and life took on a slightly different pattern, at least for Judy it did, with two children to get to and from school.

We no longer worried about the rival garden centre, as it did not seem to make any inroads into our business and ours was getting better as time went on. What could possibly upset our idyllic lifestyle? We soon found out.

Coming home from work one afternoon, the place seemed oddly quiet. Although it was a beautiful day, the kids were not outside as they normally would be and when I entered the house not even the dog was there to greet me. Where could they be?

Calling out 'Hello!' I went into the kitchen, to find Judy huddled in a chair and the children and dog gazing mournfully at her. What could be wrong?

My greatest fears were to be confirmed. She had hurt her back and could hardly move because of the pain. Undeterred by this she had somehow dragged herself down to school to collect Nigel and Catherine. What had she done?

As I have said before, Judy is very enthusiastic in everything she does. We had warned her before about not taking on some of the heavier tasks in the garden and told her until we were blue in the face to either get someone else capable to do the job, or to wait for Cliff or me to turn up.

We had had a delivery of peat bales and other sundry items and as usual, she had ignored our entreaties to leave it to the delivery driver to unload and had been heaving the peat bales into the garage. These were the large one-hundredweight type and even took Cliff and I some muscle power to move. She had probably pulled a muscle or even worse, got a slipped disc.

Our part-time helper was pressed into service to cover whilst we found out. On visiting her doctor and having an X-ray, we discovered our concern was justified, she had a slipped disc. She was fitted up with what we called a cast-iron corset but was really assembled from heavy canvas on a steel frame. It was very cumbersome but it did provide some relief.

Soon, we seemed to be back to normal, except that when Judy got a sale of a tree or something heavy, she could not lift it. Customers seemed strangely incapable or unwilling to carry anything themselves, even quite beefy looking men. When I got home, I was faced with not only lifting and carrying the item up to the drive, but having to deliver it as well. This naturally bit into our profits. We even lost money if the item had to be delivered a long distance. We could not go on like this for very long. We were reluctant to increase prices, as this had been our main bulwark against our competitors.

We investigated the possibility of Cliff or I taking over full-time but this would have necessitated a large expansion of the business, which in turn needed capital, and this we did not have. We were struggling to pay off our loan as it was. At the end of the year when we would be normally ordering our next year's stock, we decided that enough was enough. Sadly, we would have to call it a day.

Chapter Forty-Five

1975 was not a good year. We could only dwell on the good times and even the bad times that had been. We now had to make the best of a bad job. So, putting on a brave face, we set about selling off our stock for the best price we could get.

To start with, we tried to keep up the pretence that things were normal but, as we could no longer supply items we would normally have for sale, this proved very difficult. We did buy in the usual bedding plants – as we knew that we would have no trouble selling them – but after that we stopped buying.

We put everything up for sale, advertising it as a closing down sale, which is what it reluctantly, was. It was quite amazing that former customers, who would not have been seen dead carrying anything so heavy as a tree, suddenly developed muscles capable of whisking their purchases up to the drive and into their vehicles, which in some cases they had not been prepared to sully with dirty items before. At weekends it was almost as though a swarm of locusts had moved in. We could hardly keep up with customers thrusting money into our hands.

It was also obvious that once the Garden Centre was closed, we would not be able to keep the house and grounds. Several customers asked what we would be doing with it and we had to admit that we would be forced to sell it. Although a number expressed an interest, none came forward with an offer.

It would be a terrible wrench, especially to Judy, whose heart would be broken by having to leave a place she had made home for thirteen years. This would also be true for Nigel and Catherine, who had known no other home.

Now that we had finally decided that we should sell the property, we looked at the options. No one was likely to buy it to continue using it as a garden centre, due to the cost of upkeep of the house. We had our full-time jobs and people in a similar position were going to be few and far between. What would help its sale?

New housing had moved into the area as we had predicted and with access to our land at the back from the side road, there seemed to be an opportunity to use part of it for residential development. Being in the profession, I was well placed to make an application for outline planning permission on the bottomland and paddock.

We were pleasantly surprised to receive permission without any problems. We felt that although it was unfortunate that this would split up the estate, more pragmatically, it would enhance the price and prospect of a sale; we needed every penny to buy a decent property. Moving down from a six-bedroom house with two acres of ground would be traumatic. Where did one put the furniture? How would we deal with less garden space?

We approached a prominent estate agent who dealt in large properties and stressed the need to sell either the property with all the land or the property with a smaller part of the land. We did not want the land with housing permission sold before the house or we would be left with a house overlooking an estate, which was not going to be very saleable.

Unbelievably, all the people sent round to view were only interested in the building land, not the house. We also had a number of offers through the agents, again for the land only.

I went round to the agents to reinforce the fact that it was the house we were more interested in selling. No one listened. More offers for the land came in from the agents. This was not on. We went round again to the agents and much to their surprise, withdrew the property from sale. What did they expect? They were not carrying out our wishes, only looking for a fast buck on a sale.

What now? Should we approach another estate agent or were we just saddled with an unsaleable property? We did consider splitting the house into two, but it would not have been the same living in only part of it, even though we used less than half at the present time.

In the meantime, we continued the rundown of the nursery stock. This brought in large numbers of people and inevitably, some asked what we were going to do with the house and land.

We confirmed that it was up for sale but, although several people expressed an interest, most used it as an excuse to just nose around.

Finally, one customer showed much more dedication, coming back twice more to view. He was not interested in the building land he was more keen on acquiring the house for use as a licensed restaurant, with the possibility of building on an hotel wing with ancillary sports facilities, such as squash courts.

He agreed to purchase, subject to getting permission for these additions. So it was back to obtaining an outline permission for these new proposals. It was just as well I was in the right profession, otherwise it would have cost us a fortune. Once again, we were lucky and got clearance through, not without a lot of crossed fingers and prayers. Would he renege on the agreement?

Luckily, he was very positive and we agreed a purchase price, which was ten times what we had paid for it thirteen years ago. Mind you, house prices in general had gone up.

Panic now set in. We had not made any attempt to look for anything for ourselves. It had been so difficult finding a buyer for our kind of property, we did not want to be disappointed if we had found something that we wanted before we had managed to sell. The purchaser was very accommodating, saying we could have time to find something we liked, although we realised that this could not be taken too far.

We visited every estate agent in the area who had property for sale in Glossop and the surrounding areas, as we had decided to stay. We liked it but more than that, our children were settled with school and friends. What we wanted, was a largish four-bedroom detached house, with a reasonable amount of land. This ruled out new property, as the gardens of these were hardly big enough to swing a cat.

We went to a number that sounded promising from the agent's blurb. How they had the cheek to describe some of them the way they did was beyond belief. One was described as having large grounds with extensive views, snuggling in a small village of character. It turned out to be trapped between the telephone exchange and a small factory with a council estate at the back and a bus stop right outside the front bedroom windows. The photographer must have been a contortionist to take the advertising shot.

Others were hardly much better and after five weeks of searching, we were beginning to despair. Naturally, our friends and acquaintances were aware of what we were looking for and were keeping their eyes open for any likely properties.

Then, one day, one of our friends phoned up to say that he had just passed a house in the nearby village of Simmondley where a workman was putting up a 'For Sale' board. Simmondley was just the other side of the railway line from Gamesley House. We thanked him, and promised to investigate. Almost immediately, there was a knock on the door and there stood our former helper. He told us that he had just passed this house in Simmondley. Isn't the coincidence amazing?

We quickly drove round to investigate from the outside. It was stone-built on a corner site of about half an acre with a stream running through. We did not seem to be able to get away from water. (In fact, our present home has a one-acre lake at the bottom of the garden.)

The agents were out of the area, some eight miles away, so I set out to get the particulars. It proved to be only a three-bed house, but the main bedroom was twenty-two feet long by fourteen-foot wide, which would certainly make us feel at home, comparing it with Gamesley. Would that bedroom split into two? We arranged to go and view it that weekend.

Our first impressions were favourable. That main bedroom had a doorway at the corner of two blank walls, with a washbasin on the back wall. There were a group of five mullioned windows on the front and three on the side. I could see that it could be split up into two, with a lobby containing the washbasin.

The house was very unusual in that it was shaped like an off-centre cross, which meant that all rooms had three outside walls. It wasn't looking very good for heat retention, but the house had eighteen inch thick walls with stone outer leaf and this would hopefully hold the heat.

It turned out that there were two streams, one running across the front garden, which fed by a small waterfall into a lower stream between the main garden and a woodland area, bounded by the road. Most importantly, there was a sixteen-foot by eight-foot greenhouse. It did not compare to the hundred-foot of greenhouse at Gamesley, but at least there was one. There was also a larger than normal double garage, a utility room and WC attached to the rear.

Back home, we discussed its merits. I loved it, Cliff was quite happy with it, but Judy had some reservations, which she could not define, just a feeling.

Feelings or no feelings, it was the best we had seen and time was passing by; we had to make a decision. It was two against one, but Judy, in spite of her misgivings, agreed to go along with it. We rang the vendor to go and talk about such things as carpets, curtains and so forth. We had learnt the hard way from our last attempt to purchase. Within three days, we had agreed to buy.

Now came the hassle. We could not possibly take all that furniture. We had to decide what would fit and leave or sell the rest. Moving down from such a large house was not easily done. It was amazing how much you manage to accumulate when you have the space. Finally, everything was set to move, including getting a builder ready to split that bedroom, together with some other small jobs.

When did we move? Once again in midwinter, this time on Cliff's birthday in January. Next time we have to move, perhaps it would be nice to try on Judy's birthday in May, when we could hope for better weather.

It was a very sad family who bid farewell to that fantastic place that had been our home and business for thirteen years. At least we knew it would find another use, when so many other similar properties were either falling into disrepair or left empty.

Over the years that followed, Judy refused point blank to visit the newly set up restaurant. That had been her happy home, with all its memories. Seeing it in its new guise would not be right.

My son and daughter, with great subterfuge, did manage to get her there once for her fortieth birthday. The owner gave us a conducted tour, but it felt alien and we never went again. The final chapter of this part of our lives had come to an end. No more would the answer lie in the soil.

Well, not quite! Our new home had a half-acre garden and we

could not get out of the habit of growing hundreds of seedlings. Luckily, that chap who had started up a nursery against us sold out and we became quite friendly with the new owners. We effectively bartered our excess plants for the garden wares we required, saving us quite a lot of money.

Since then we have moved to a further property in Glossop, which also has half an acre of garden. It was established in 1930 but over the latter years had rather been neglected. We are hard at work completely transforming it into a garden we can be proud of and which will serve us in our fading years.

We still have our happy memories, particularly those amusing episodes recounted here and we are still producing too many seedlings!